# CRISPR-Cas System in Translational Biotechnology

## Rickbed Nandi

# Copyright

## Disclaimer

This academic book, titled "**CRISPR-Cas System in Translational Biotechnology**," is published in the year 2023 and represents the culmination of extensive research and scholarship by the author in their respective fields of expertise. As readers engage with the content presented within these pages, we would like to provide the following disclaimer:

***Accuracy and Currency***: While every effort has been made to ensure the accuracy and currency of the information contained in this book, the dynamic nature of academic disciplines and the ever-evolving body of knowledge may result in some content becoming outdated over time. Readers are encouraged to verify the information presented herein with the latest research and developments in their respective fields.

***Authorship and Perspectives***: This book may reflect the views, opinions, and interpretations of the author based on their research and expertise. The author does not claim to represent a single, universally accepted perspective on the subject matter. The diversity of thought within academic disciplines is both acknowledged and respected.

***Citations and References***: Proper citations and references have been provided to acknowledge the sources of ideas, data, and information used in this book. Readers are encouraged to consult the cited sources for in-depth exploration and to verify the accuracy of referenced material.

***Limitation of Liability***: The author, publisher, and any affiliated institutions or individuals associated with this book disclaim any liability for any errors, omissions, or damages that may result from the use or interpretation of the information presented herein. Readers are encouraged to exercise critical thinking and academic discernment when engaging with the content.

*Ethical Considerations*: The author has made every effort to adhere to ethical guidelines in their research and writing. However, the responsibility for ethical considerations, such as proper attribution and adherence to ethical research practices, ultimately rests with the individual researcher or reader.

*Copyright*: All rights, including but not limited to the rights of reproduction, distribution, and translation, are reserved. No part of this book may be reproduced or transmitted in any form or by any means without prior written permission from the publisher, except for brief quotations in critical reviews and scholarly works.

*Reader Responsibility*: Readers are encouraged to engage with this book critically and responsibly. This includes evaluating the relevance of the content to their specific academic pursuits, as well as respecting copyright and intellectual property rights when quoting or referencing this work.

Finally, this book is offered as a contribution to the academic community and the broader intellectual discourse. It is our hope that it serves as a valuable resource for scholars, students, and anyone interested in the subject matter. However, readers are reminded that academic knowledge is continually evolving, and they should supplement their understanding with the latest research and developments in their respective fields.

## Preface

Welcome to the world of "*CRISPR-Cas System in Translational Biotechnology*," a book that explores the cutting-edge applications and implications of one of the most revolutionary technological advancements in modern science. In these pages, we embark on a journey through the dynamic landscape of CRISPR-Cas technology, guided by the expertise and insights of numerous researchers, experts, and visionaries in the field.

The CRISPR-Cas system, with its humble origins as a bacterial immune system, has evolved into a versatile and powerful tool that has the potential to reshape the way we approach biotechnology, medicine, agriculture, and beyond. As a writer and editor, I have had the privilege of witnessing the rapid growth and transformation of this field, and it is with great enthusiasm that I share this comprehensive exploration of CRISPR-Cas with you.

This book is a culmination of years of dedicated research, collaboration, and innovation by scientists, engineers, and thought leaders from around the world. It is designed to provide a deep understanding of the CRISPR-Cas system and its myriad applications in translational biotechnology. Each chapter delves into a specific aspect of CRISPR-Cas technology, from its fundamental molecular mechanisms to its transformative impact on fields such as genome editing, disease modelling, agriculture, drug discovery, and more.

But our journey does not stop at the laboratory bench. We also examine the ethical, legal, and societal dimensions of CRISPR-Cas technology, acknowledging the profound responsibility that comes with such powerful tools. We contemplate the opportunities and challenges that lie ahead as we navigate the ever-evolving landscape of biotechnology, striving for a future where CRISPR-Cas serves humanity in a responsible and ethical manner.

I hope this book serves as both a valuable resource and an inspiration for students, researchers, healthcare professionals, policymakers, and anyone interested in the remarkable world of CRISPR-Cas technology. It is my sincere belief that through understanding and responsible stewardship, we can harness the potential of CRISPR-Cas for the betterment of society.

I would like to express my heartfelt gratitude to the contributors who have shared their expertise and passion, as well as to the readers who embark on this journey with us. The story of CRISPR-Cas is still being written, and I am excited to have you as part of this remarkable narrative.

Sincerely,

Rickbed Nandi

## Dedication

To the Curious Minds and Inquisitive Spirits,

This book is dedicated to the general readers and academic community, whose insatiable thirst for knowledge and unwavering commitment to the pursuit of scientific excellence have been the driving force behind the incredible advancements in the field of CRISPR-Cas technology.

In your collective curiosity, you have sparked revolutions in genome editing, disease modelling, agriculture, and beyond. Your passion for discovery has illuminated the path toward a brighter, more sustainable future for all of humanity. Your dedication to the pursuit of truth, ethics, and innovation has elevated the CRISPR-Cas system from a scientific breakthrough to a beacon of hope for generations to come.

May this dedication stand as a testament to your boundless curiosity, relentless dedication, and enduring commitment to the advancement of knowledge. With heartfelt gratitude, we offer this book as a humble contribution to your journey, in the hope that it may inspire new horizons and spark even greater discoveries in the ever-expanding realm of translational biotechnology.

With admiration and respect,

Rickbed Nandi

Tapasi Mandal

# Contents

# Chapter 1: Introduction to CRISPR-Cas System

## 1.1 The Discovery of CRISPR-Cas

The discovery of the CRISPR-Cas system is a fascinating tale of scientific curiosity, international collaboration, and the power of basic research to revolutionize biotechnology and molecular biology. In this subsection, we delve into the historical timeline and key players who contributed to unravelling the mysteries of CRISPR.

### Early Observations and Hints at CRISPR

The story of CRISPR's discovery can be traced back to the early 1980s when Japanese researchers stumbled upon a peculiar set of repeating DNA sequences in the genome of the bacterium Escherichia coli (E. coli). However, these repetitive sequences, initially known as "short regularly spaced repeats" or SRSRs, remained largely enigmatic for several years.

In the late 1980s and early 1990s, similar repetitive patterns were observed in the genomes of other bacteria and archaea, but their function remained elusive. It wasn't until the late 1990s that the breakthrough came when researchers found that these repeats were often associated with "spacer" sequences - unique DNA sequences of viral origin.

### The Dawn of Comparative Genomics

The turning point in understanding the significance of these repeats and spacers came with the advent of comparative genomics. In 2002, Francisco Mojica, a Spanish microbiologist, made a pivotal observation. He noticed that these repeats and spacers were remarkably conserved across different strains of bacteria and archaea. This conservation hinted at a potentially critical biological function.

## CRISPR Terminology and Classification

Mojica's work prompted the adoption of the term "CRISPR" (Clustered Regularly Interspaced Short Palindromic Repeats) to describe these repetitive elements. Subsequently, the associated proteins that were thought to be involved in the system were named "CRISPR-associated" or "Cas" proteins.

CRISPRs were initially classified into two main types based on their gene content: Type I and Type II. This classification was later expanded to include additional types and subtypes, each characterized by unique Cas proteins and mechanisms. The discovery of these different types hinted at the diverse functionalities of the CRISPR-Cas systems.

## International Collaboration and the Role of Yogurt

One of the critical breakthroughs in understanding CRISPR-Cas systems came from a rather unexpected source: yogurt. In 2005, researchers working at the Danish food company Danisco (now part of DuPont) were investigating the fermentation process used in yogurt production. They were specifically interested in Streptococcus thermophilus, a bacterium used in yogurt fermentation.

During their research, they noted the presence of CRISPR-Cas systems in the genome of S. thermophilus. This revelation marked a significant milestone, as it demonstrated that CRISPR-Cas systems were not limited to pathogens but could also be found in beneficial bacteria.

## The Role of CRISPR in Bacterial Immunity

Around the same time, researchers Marraffini and Sontheimer, working at Northwestern University, made a groundbreaking discovery. They found that the CRISPR-Cas system in

Streptococcus pyogenes (a bacterium responsible for strep throat and other infections) played a crucial role in protecting the bacterium from invading viruses.

This discovery shed light on the primary function of CRISPR-Cas systems in bacteria: they function as an adaptive immune system. Bacteria capture and store genetic material from viruses they've encountered in the past as "spacer" sequences. When a virus attacks again, the bacteria can use these spacers as a template to produce RNA molecules that target and destroy the virus's genetic material.

## The CRISPR-Cas9 Revolution

While the early discoveries of CRISPR-Cas were monumental, it was the breakthrough involving the Cas9 protein that truly catapulted CRISPR into the limelight. Jennifer Doudna and Emmanuelle Charpentier, two prominent scientists, collaborated to decipher the molecular mechanisms behind the CRISPR-Cas9 system.

In 2012, Doudna and Charpentier published a landmark paper detailing how the Cas9 protein, guided by a synthetic RNA molecule, could be programmed to precisely target and cleave specific DNA sequences. This discovery unlocked the potential for precise and easy genome editing, marking a revolution in biotechnology.

## Applications and Implications

The discovery of CRISPR-Cas9 set off a flurry of research and applications across various fields. Within a few years, the system was being used for gene editing in a wide range of organisms, from bacteria to plants, animals, and even humans. Its applications extended to disease modelling, drug development, and potential treatments for genetic disorders.

However, the power of CRISPR-Cas9 also raised important ethical and regulatory questions. The ability to edit the human germline and the potential for unintended consequences sparked debates about responsible use and the need for stringent oversight.

The discovery of the CRISPR-Cas system was a journey marked by curiosity, collaboration, and scientific perseverance. From the initial observations of repetitive DNA sequences to the development of CRISPR-Cas9 as a genome-editing tool, the story of CRISPR is a testament to the capacity of science to unravel the mysteries of nature and shape the future of biotechnology.

## 1.2 Evolution and Diversity of CRISPR Systems

The evolution and diversity of CRISPR systems are central to understanding the remarkable range of functions this adaptive immune system serves in prokaryotes. The discovery of CRISPR-Cas systems has opened up a new frontier in microbiology, offering insights into how microorganisms defend against viral invaders and adapt to their ever-changing environment. In this subsection, we will delve into the intriguing evolutionary history and the fascinating diversity of CRISPR systems.

### The Origins of CRISPR: An Ancient Battle Against Viruses

The story of CRISPR-Cas system evolution dates back billions of years, making it one of the most ancient defence mechanisms in the microbial world. It all begins with the perpetual battle between bacteria and viruses, known as phages. To survive in this ongoing war, bacteria have developed an arsenal of defence mechanisms, with CRISPR-Cas being one of the most sophisticated.

One of the early clues to the existence of CRISPR-Cas systems came from the observation that some prokaryotic genomes contained peculiar repetitive sequences interspersed with unique sequences of viral origin. These repetitive elements were aptly named Clustered Regularly Interspaced Short Palindromic Repeats or CRISPR. The unique sequences between the repeats were identified as "spacers," and it was soon realized that these spacers matched the genetic material of viruses and plasmids encountered by the bacteria.

Researchers speculated that CRISPRs might play a role in adaptive immunity, allowing bacteria to store genetic information from past viral encounters and use it to fend off future attacks. This hypothesis was later confirmed, leading to the revelation that CRISPRs, in combination with Cas proteins, constitute a potent immune system that can be harnessed for various biotechnological applications.

## Diversity of CRISPR-Cas Systems: A Taxonomic Treasure Trove

The diversity of CRISPR-Cas systems is nothing short of astonishing. Over time, researchers have identified multiple types and subtypes of CRISPR-Cas systems, each with its unique features and functionalities. This diversity is largely attributed to the co-evolution of bacteria and their viral adversaries. Different bacteria have adapted CRISPR-Cas systems to recognize and combat a wide range of viruses and mobile genetic elements.

### Classification of CRISPR-Cas Systems

To categorize this diversity systematically, scientists have developed a classification scheme based on the signature Cas proteins found in each system. The two main classes of CRISPR-Cas systems are Class 1 and Class 2, each with several types and subtypes.

*Class 1 CRISPR-Cas Systems*: These systems are characterized by their multi-protein complexes, often involving several Cas proteins working together. Examples include Type I and Type III systems. In Type I systems, for instance, the Cas proteins form a complex known as Cascade (CRISPR-Associated Complex for Antiviral Defence).

*Class 2 CRISPR-Cas Systems*: Class 2 systems are characterized by a single, large Cas protein that is responsible for both target recognition and interference. The most well-known Class 2 system is Type II, which includes the Cas9 protein.

Within each class, there are further subtypes, and each subtype exhibits unique features regarding target recognition, interference mechanisms, and adaptation. For instance, the Type II CRISPR-Cas9 system has been widely used for genome editing due to its simplicity and precision.

### Uncovering New CRISPR Systems

The diversity of CRISPR-Cas systems is not limited to the known classes and subtypes. Ongoing research continues to uncover novel systems, expanding our understanding of prokaryotic immune strategies. Some newly discovered CRISPR-Cas systems challenge our preconceptions about the limits of these systems' capabilities.

One such example is the Type V CRISPR-Cas12 system, which has been dubbed "Cpf1" after its signature Cas protein. Cpf1 has unique properties compared to Cas9, including the ability to recognize different protospacer adjacent motifs (PAMs) and generate staggered double-stranded DNA breaks. This versatility has broadened the toolkit available for genome editing and gene regulation.

## The Role of CRISPR-Cas System Diversity in Microbial Communities

Beyond its utility in biotechnology, the diversity of CRISPR-Cas systems plays a critical role in shaping microbial ecosystems. These systems contribute to the ecological balance by influencing the interactions between bacteria and their viral predators.

In microbial communities, different bacteria may possess distinct CRISPR-Cas systems that are tailored to combat specific phages. This specialization allows for niche partitioning, reducing competition among bacterial species. Furthermore, horizontal gene transfer (HGT) of CRISPR-Cas systems between bacteria can influence the spread of immunity strategies throughout microbial populations.

## Implications for Translational Biotechnology

Understanding the evolution and diversity of CRISPR-Cas systems is vital for harnessing their full potential in translational biotechnology. Researchers are continually exploring novel CRISPR systems and developing new tools based on their unique properties. This diversity not only provides a broader range of applications but also offers solutions to specific challenges posed by different genetic targets.

The evolution and diversity of CRISPR-Cas systems underscore the complexity and adaptability of prokaryotic immune systems. These systems have evolved over billions of years to defend against viral invaders, and their remarkable diversity continues to drive innovation in biotechnology and expand our understanding of microbial ecosystems.

## 1.3 CRISPR-Cas Classification and Types

The CRISPR-Cas system, a revolutionary genome editing tool, is not a one-size-fits-all technology. Instead, it comprises a diverse array of molecular machinery tailored for different tasks. CRISPR-Cas systems are classified into two primary classes (Class 1 and Class 2) and further subdivided into types and subtypes based on their signature proteins and mechanisms. Understanding this classification is crucial for harnessing the full potential of CRISPR-Cas in translational biotechnology.

### Class 1 CRISPR-Cas Systems

Class 1 CRISPR-Cas systems are characterized by their multi-protein complexes, in contrast to the single-protein systems found in Class 2. These systems are more complex but offer unique advantages.

### Type I CRISPR-Cas Systems

Type I CRISPR-Cas systems are the largest and most complex class within Class 1. They are commonly found in bacteria and archaea. These systems are defined by the presence of the signature protein, Cas3, which is involved in both interference and adaptation phases. Type I systems use a Cascade complex for interference, consisting of several Cas proteins (Csm or Cmr), and a crRNA molecule. The Cascade complex recognizes target DNA and degrades it, usually with the assistance of Cas3.

*Example*: A study by Jackson et al. in 2017, published in "Nature," explored the application of Type I CRISPR-Cas systems for multiplex genome editing in E. coli. They demonstrated the simultaneous targeting of multiple genes using the Cascade complex, highlighting its potential in synthetic biology.

## Type III CRISPR-Cas Systems

Type III CRISPR-Cas systems are another Class 1 subtype found primarily in prokaryotes. Unlike Type I, Type III systems use the signature protein, Csm or Cmr, for both interference and adaptation. These systems are known for their cyclic transcription-dependent processing (CTD) mechanism.

*Example*: A study by Samai et al. in 2015, published in "Molecular Cell," elucidated the CTD mechanism in Type III-A CRISPR-Cas systems of Sulfolobus solfataricus. Their findings shed light on the molecular intricacies of Type III systems, facilitating their potential applications.

## Type IV CRISPR-Cas Systems

Type IV CRISPR-Cas systems are the newest addition to Class 1, and research on them is ongoing. They are characterized by the presence of Csf1, a signature protein, and they are found in both bacteria and archaea.

*Example*: While there isn't an established example of Type IV CRISPR-Cas applications at this stage due to their recent discovery, ongoing research in extremophiles and environmental samples is expected to uncover their unique capabilities.

## Class 2 CRISPR-Cas Systems

Class 2 CRISPR-Cas systems are simpler in structure compared to Class 1 and have been the primary focus of genome editing applications due to their compact nature.

## Type II CRISPR-Cas Systems

Type II CRISPR-Cas systems are the most well-known and extensively used class within Class 2. They are characterized by the signature protein, Cas9, and a single guide RNA (sgRNA) molecule that guides Cas9 to its target DNA sequence. Type II systems have been harnessed for precise genome editing across various organisms, making them a cornerstone of CRISPR technology.

*Example*: The groundbreaking work of Doudna and Charpentier in 2012, published in "Science," detailed the development of the CRISPR-Cas9 system for genome editing. This discovery has revolutionized the field of biotechnology and medicine.

### Type V CRISPR-Cas Systems (Cpf1/Cas12)

Type V CRISPR-Cas systems, also known as Cpf1 or Cas12, are emerging as powerful genome editing tools. They are characterized by the signature protein, Cpf1, and require a single CRISPR RNA (crRNA) molecule for target recognition. One notable feature of Cpf1 is its ability to process its own crRNA.

*Example*: In a study by Zetsche et al. in 2015, published in "Cell," the authors described the development of Cpf1-based genome editing. They demonstrated its efficiency in targeting human cells and its potential as an alternative to Cas9 for precise genome modifications.

### Type VI CRISPR-Cas Systems (C2c2/Cas13)

Type VI CRISPR-Cas systems, or C2c2/Cas13, have gained attention for their unique RNA-targeting capabilities. The signature protein, C2c2 (now known as Cas13), has RNase activity and can be programmed to target specific RNA sequences, making it a valuable tool for RNA manipulation and diagnostics.

*Example*: Gootenberg et al., in a study published in "Science" in 2017, introduced the concept of using Cas13 for specific RNA detection, termed SHERLOCK (Specific High-sensitivity Enzymatic Reporter UnLOCKing). This technology has applications in diagnosing infectious diseases and detecting RNA viruses.

The classification of CRISPR-Cas systems into Class 1 and Class 2, along with their respective types and subtypes, underscores the diversity and adaptability of this revolutionary technology. Each class and type offers unique advantages and potential applications in translational biotechnology, from genome editing to diagnostics and beyond. Researchers continue to explore and unlock the full potential of CRISPR-Cas systems, leading to exciting developments in the field.

# Chapter 2: Molecular Mechanisms of CRISPR-Cas

## 2.1 CRISPR Adaptation and Spacer Acquisition

The CRISPR-Cas system, known for its remarkable genome editing capabilities, has its origins in a fundamental biological process called "CRISPR adaptation." This process involves the acquisition of genetic material, known as "spacers," from invading viruses and plasmids and their incorporation into the host organism's CRISPR array. Understanding the intricacies of CRISPR adaptation is pivotal for harnessing the full potential of this technology in translational biotechnology. In this section, we delve into the mechanisms and significance of CRISPR adaptation, backed by relevant examples and scientific data.

## Mechanisms of CRISPR Adaptation

CRISPR adaptation begins when a host organism encounters a foreign genetic element, such as a virus or plasmid. The organism recognizes the invader's DNA or RNA and initiates a defence response that culminates in the integration of a small segment of the invader's genetic material into the CRISPR array as a spacer. This process can be divided into several key steps:

*Recognition of Invading DNA*: The initial step involves the detection of foreign DNA or RNA by the host's surveillance machinery. In the case of Type II CRISPR-Cas systems, such as CRISPR-Cas9, the Cas proteins play a critical role in identifying and binding to the foreign nucleic acids.

A study conducted by Barrangou et al. in 2007 demonstrated that Streptococcus thermophilus employs its CRISPR-Cas system to resist phage infection. The Cas proteins in this bacterium recognize and bind to the phage DNA, initiating the adaptation process.

*Processing and Capture of Spacers*: Once the foreign genetic material is recognized, it is enzymatically processed to generate small DNA fragments, known as protospacers. These protospacers are then integrated into the host's CRISPR array as new spacers.

In 2012, Datsenko et al. conducted experiments with Escherichia coli, illustrating the capture of protospacers from invading plasmids. This research highlighted the role of Cas proteins, such as Cas1 and Cas2, in spacer acquisition.

*Integration into the CRISPR Array*: The newly acquired spacers are inserted into the CRISPR array, typically located within the host's genome, flanked by repeat sequences. This integration is catalysed by Cas1 and Cas2 or other associated proteins.

A study published by Yosef et al. in 2012 detailed the integration of spacers into the CRISPR array of the bacterium Vibrio cholerae. This research provided insights into the integration process and the role of accessory factors.

*Adaptation Memory*: The CRISPR array now contains a record of past encounters with invaders, forming the basis of the adaptive immune system. When the same or a similar invader attacks again, the CRISPR-Cas system can target and destroy it more effectively.

## Significance of CRISPR Adaptation

CRISPR adaptation is not just a fascinating biological phenomenon; it also has profound implications for various applications in translational biotechnology:

*Improved Genome Editing Specificity*: The spacers acquired during CRISPR adaptation serve as a memory bank of past encounters. By designing guide RNAs that target specific sequences found in these spacers, researchers can enhance the specificity of genome editing, reducing off-target effects.

A study published in Nature Biotechnology in 2018 by Kim et al. demonstrated how the use of spacers acquired through adaptation can improve the specificity of CRISPR-Cas9 gene editing, reducing unintended mutations.

*Disease Diagnostics*: CRISPR adaptation can be leveraged for the development of highly sensitive diagnostic tools. The presence of specific spacers in a CRISPR array can indicate past infections or exposures to particular pathogens.

*Example*: In a study published in PLoS ONE in 2015, researchers utilized CRISPR arrays in Mycobacterium tuberculosis to detect the presence of specific spacers associated with drug resistance genes. This approach allowed for rapid diagnosis of drug-resistant tuberculosis strains.

*Biotechnology Innovation*: CRISPR adaptation has inspired innovative biotechnological applications. For instance, engineered CRISPR-Cas systems with customized spacers can be used to detect specific DNA or RNA sequences in a highly targeted manner.

*Data*: A research article published in Science in 2019 by Gootenberg et al. described the development of the SHERLOCK (Specific High-sensitivity Enzymatic Reporter UnLOCKing) platform, which utilizes adapted CRISPR systems for nucleic acid detection with high sensitivity.

*Environmental Monitoring*: CRISPR adaptation can be employed for monitoring and studying microbial communities in various environments, including soil, water, and the human microbiome. By analysing the spacer content, researchers can gain insights into the history of microbial interactions.

*Example*: A study published in Environmental Microbiology in 2020 by Rho et al. used CRISPR adaptation analysis to explore the dynamics of microbial communities in marine environments, shedding light on their responses to changing conditions.

CRISPR adaptation is a pivotal process that underlies the functionality of the CRISPR-Cas system. Its mechanisms are increasingly well-understood, and its significance extends to a wide array of applications in translational biotechnology, from enhancing genome editing precision to innovative disease diagnostics and environmental monitoring. As our understanding of CRISPR adaptation deepens, its potential for transformative applications continues to grow.

## 2.2 CRISPR RNA Biogenesis

CRISPR RNA (crRNA) biogenesis is a crucial step in the functioning of the CRISPR-Cas system. It involves the conversion of precursor CRISPR RNAs into mature crRNAs, which guide the Cas protein machinery to target and cleave specific DNA or RNA sequences. Understanding the intricacies of crRNA biogenesis is pivotal for harnessing the full potential of CRISPR-Cas technology.

### crRNA Precursor Generation

The crRNA precursor molecules are transcribed from the CRISPR arrays, which consist of short, repetitive DNA sequences interspersed with unique spacers derived from past encounters with foreign genetic elements, such as viruses or plasmids. In bacteria and archaea, CRISPR arrays are transcribed into long precursor transcripts known as precursor CRISPR RNAs (pre-crRNAs). These pre-crRNAs serve as the primary source of crRNAs.

In the initial step of crRNA biogenesis, the pre-crRNA is transcribed by RNA polymerase. The resulting transcript is typically a long, multi-repeat RNA molecule that encompasses all the spacer sequences within the CRISPR array. For example, in the well-studied CRISPR system of Streptococcus pyogenes, the pre-crRNA transcript includes all the repeat-spacer units within the array.

Researchers studying the CRISPR-Cas system in Streptococcus pyogenes have identified that the pre-crRNA molecule in this bacterium can be several kilobases long, containing multiple repeat-spacer units (Deltcheva et al., 2011).

## crRNA Processing

The next critical step in crRNA biogenesis is the processing of the long pre-crRNA transcript into individual crRNAs, each specific to a unique spacer sequence. This processing step is essential for ensuring that the CRISPR-Cas system can accurately target and cleave invading genetic material.

In many CRISPR systems, endoribonucleases play a crucial role in crRNA processing. These endoribonucleases cleave the pre-crRNA at specific locations within the repeat sequences, generating individual crRNAs. The exact mechanisms of processing can vary between different CRISPR systems and organisms.

In the Type II CRISPR system found in Streptococcus pyogenes, the endoribonuclease Csn1 (also known as Cas9) plays a central role in processing the pre-crRNA into individual crRNAs, each containing a single spacer sequence (Jinek et al., 2012).

## crRNA Modifications

After processing, crRNAs may undergo further modifications that can influence their stability and efficacy. These modifications include the addition of chemical groups or structural alterations that fine-tune the crRNA's interaction with the Cas proteins and target nucleic acids.

One common modification is the addition of a 5' triphosphate group to the crRNA. This modification is thought to protect the crRNA from degradation by cellular exonucleases and is essential for efficient binding to the Cas protein.

Recent research has shown that the addition of a 5' triphosphate group to crRNAs is a conserved modification across various CRISPR systems and is critical for their function (Shiimori et al., 2020).

## crRNA Loading onto Cas Proteins

Once the crRNAs are generated and possibly modified, they need to be loaded onto the Cas proteins to form the functional ribonucleoprotein complexes that can search for and cleave target DNA or RNA. The loading process is highly specific and involves interactions between the crRNA and the Cas proteins, particularly the Cas endonuclease responsible for target recognition and cleavage.

In Type II CRISPR systems, such as CRISPR-Cas9, the loading of crRNA onto the Cas protein is a complex process. It involves the formation of a Cas protein-crRNA complex, which, in the case of Cas9, requires an additional trans-activating CRISPR RNA (tracrRNA) molecule for stability.

The loading of crRNA onto the Cas9 protein in Streptococcus pyogenes has been extensively studied and is a key step in the mechanism of genome editing using the CRISPR-Cas9 system (Jinek et al., 2012).

## crRNA Maturation and Stability

Once crRNA is loaded onto the Cas protein, it undergoes further maturation steps to ensure its stability and functionality during target recognition and cleavage. These maturation steps can include additional RNA processing or conformational changes that optimize the crRNA-Cas protein interaction.

Studies have shown that in some CRISPR systems, such as Type I CRISPR-Cas systems, crRNAs may undergo structural changes upon binding to the Cas protein, which are essential for efficient target recognition (Rouillon et al., 2013).

## Species-Specific Variations in crRNA Biogenesis

It's important to note that the process of crRNA biogenesis can vary between different CRISPR systems and organisms. Species-specific variations in the enzymes involved, the processing sites within the pre-crRNA, and the presence of additional cofactors can influence the efficiency and specificity of crRNA production.

The Type VI CRISPR-Cas system, also known as C2c2, found in some bacteria and archaea, utilizes a different set of enzymes and mechanisms for crRNA biogenesis compared to the more well-known Type II systems, highlighting the diversity in CRISPR-Cas systems (Abudayyeh et al., 2016).

CRISPR RNA biogenesis is a highly regulated and precise process that results in the generation of mature crRNAs. These crRNAs serve as the guiding molecules for the Cas protein machinery, allowing for the specific targeting and cleavage of invading genetic material. Understanding the nuances of crRNA biogenesis is essential for optimizing CRISPR-Cas technology for various applications, from genome editing to diagnostics and beyond. Additionally, ongoing research continues to unveil the diversity of CRISPR-Cas systems and the variations in crRNA biogenesis among different organisms, further expanding our knowledge of this groundbreaking technology.

## 2.3 Target Recognition and Interference

The CRISPR-Cas system, renowned for its versatile applications in genome editing and beyond, relies fundamentally on its ability to recognize and interfere with specific target sequences. This section delves into the intricate mechanisms behind target recognition and interference, highlighting key factors and breakthroughs in this aspect of the system.

### Protospacer Adjacent Motif (PAM) Recognition

One of the defining features of the CRISPR-Cas system is its reliance on a Protospacer Adjacent Motif (PAM) for target recognition. PAM is a short, conserved DNA sequence, typically 2-6 base pairs long, located immediately adjacent to the target DNA sequence. It serves as a crucial signal for the Cas proteins to distinguish foreign DNA from the host genome.

*In the widely used CRISPR-Cas9 system, the Streptococcus pyogenes Cas9 protein recognizes a PAM sequence of "NGG" (where "N" represents any nucleotide) in the target DNA. This PAM recognition step ensures that Cas9 only binds to and cleaves DNA with the correct PAM sequence, minimizing off-target effects (Jinek et al., 2012).*

PAM recognition is a fundamental safeguard against accidental targeting of the host genome, as it ensures that the CRISPR-Cas system remains highly specific to its intended target.

### Protospacer Integration and Spacer Acquisition

Before the CRISPR-Cas system can recognize a target, it must acquire a matching spacer sequence. Spacers are short DNA sequences derived from previous encounters with foreign genetic material and serve as a "memory" of past infections. The acquisition of new spacers is a crucial aspect of the adaptive immune response conferred by CRISPR-Cas.

*In the Type I CRISPR-Cas system, which includes the Cascade complex and Cas3, the acquisition of new spacers begins with the recognition of foreign DNA. Once recognized, a short fragment of the foreign DNA, known as a protospacer, is integrated into the CRISPR array as a new spacer. This process allows the CRISPR-Cas system to remember and target the invader in subsequent encounters (Barrangou et al., 2007).*

Spacer acquisition is a dynamic process that enables the CRISPR-Cas system to continually adapt to evolving threats, making it a powerful defence mechanism in prokaryotic organisms.

## Target DNA Recognition by CRISPR RNA (crRNA)

In the CRISPR-Cas system, the recognition of a specific target DNA sequence is mediated by CRISPR RNA (crRNA), which is derived from the CRISPR array and is complementary to the spacer sequence. The crRNA, often guided by associated Cas proteins, forms a complex with the target DNA to initiate interference.

*In the Type II CRISPR-Cas system, Cas9 protein forms a complex with a guide RNA (gRNA), which is engineered to be complementary to the target DNA sequence. The gRNA directs the Cas9 protein to the specific DNA target sequence, enabling Cas9 to introduce double-stranded breaks in the DNA (Jinek et al., 2012).*

This precise recognition mechanism is a key factor in the specificity of CRISPR-Cas genome editing and interference, as it ensures that the system only targets the intended DNA sequences.

## Interference Mechanisms: Cleavage and Beyond

Upon successful recognition of the target DNA sequence, the CRISPR-Cas system initiates interference mechanisms, which can vary depending on the type and subtype of the system. The most well-known interference mechanism involves the cleavage of the target DNA, rendering it nonfunctional. However, recent advances have expanded the repertoire of CRISPR-Cas interference strategies.

*In the Type VI CRISPR-Cas system, also known as C2c2, Cas13 protein targets RNA sequences rather than DNA. Upon recognition of a complementary RNA sequence, Cas13 becomes activated and can degrade both the target RNA and nearby RNA molecules, offering potential applications in RNA manipulation and diagnostics (Abudayyeh et al., 2016).*

Furthermore, in some cases, the interference mechanism may not involve direct cleavage but rather the modification or silencing of the target DNA or RNA, expanding the versatility of CRISPR-Cas systems in gene regulation and epigenetic modifications.

## Off-Target Effects and Enhancing Specificity

Despite its precision, the CRISPR-Cas system is not immune to off-target effects, where unintended DNA sequences may be cleaved or modified. Researchers have made significant strides in improving specificity to minimize these off-target effects, a critical consideration in therapeutic applications.

*Various strategies, such as the use of high-fidelity Cas proteins, modified guide RNAs, and advanced computational algorithms, have been developed to enhance the specificity of CRISPR-Cas systems. These approaches have significantly reduced off-target effects, making the technology safer and more reliable for clinical applications (Slaymaker et al., 2016).*

Achieving a balance between high specificity and efficient target recognition is an ongoing challenge in the field of CRISPR-Cas research.

## CRISPR Interference Beyond Bacteria and Archaea

While originally discovered as a prokaryotic defence mechanism, the CRISPR-Cas system has been adapted for use in a wide range of organisms, including eukaryotes. This expansion of CRISPR-Cas technology beyond its natural hosts has opened up new possibilities for genome editing and other applications.

*In eukaryotic cells, the CRISPR-Cas system has been harnessed to perform precise genome editing. Researchers have successfully used CRISPR-Cas systems to edit the genomes of plants, animals, and even humans, revolutionizing fields such as agriculture and medicine (Doudna and Charpentier, 2014).*

This extension of CRISPR-Cas technology illustrates its adaptability and broad utility across different domains of biology.

The recognition and interference mechanisms of the CRISPR-Cas system are essential components that underpin its versatility in translational biotechnology. From PAM recognition to interference strategies and specificity enhancements, understanding these mechanisms is crucial for harnessing the full potential of CRISPR-Cas in various applications, from genome editing to disease therapy and beyond.

## 2.4 The Role of Cas Proteins

The CRISPR-Cas system is a powerful and versatile tool in genome editing and gene regulation, and at the heart of this system are Cas proteins. Cas proteins are the molecular workhorses responsible for the precision and specificity of CRISPR-based technologies. In this subsection, we will explore the various roles of Cas proteins in the CRISPR-Cas system, their structural diversity, and their applications in translational biotechnology.

### Structural Diversity of Cas Proteins

Cas proteins encompass a diverse group of endonucleases, helicases, and RNA-binding proteins, each with a unique role in the CRISPR-Cas system. One of the most well-known Cas proteins is Cas9, which plays a central role in the widely used CRISPR-Cas9 genome editing system. Cas9 is an endonuclease with two distinct domains: the HNH domain and the RuvC-like domain. These domains are responsible for cleaving the DNA strands in a sequence-specific manner. The guide RNA (gRNA) directs Cas9 to the target DNA sequence, where it forms a complex and introduces double-strand breaks (DSBs) at the desired site.

Another important class of Cas proteins includes Cas12 and Cas13. Cas12, also known as Cpf1, is an endonuclease that exhibits unique properties, such as the ability to generate staggered DNA cuts and the capacity to recognize T-rich PAM (protospacer adjacent motif) sequences, unlike the G-rich PAM recognized by Cas9. Cas12 has gained attention for its potential applications in genome editing and diagnostic assays.

Cas13, on the other hand, is an RNA-targeting endonuclease with collateral RNA cleavage activity. It has been harnessed for RNA-specific editing and the development of RNA detection tools. The collateral cleavage activity of Cas13 has been leveraged in diagnostic tests, allowing the specific detection of RNA viruses like SARS-CoV-2.

## Role of Cas Proteins in Target Recognition and Cleavage

Cas proteins play a pivotal role in the CRISPR-Cas system's ability to recognize and cleave specific DNA or RNA sequences. This recognition is facilitated by the guide RNA (gRNA), which guides the Cas protein to the complementary target sequence. The gRNA is a synthetic molecule engineered to contain a customizable 20-nucleotide sequence that matches the target sequence. This high degree of customization makes the CRISPR-Cas system versatile and adaptable to various applications.

Upon binding to the target DNA or RNA, Cas proteins initiate cleavage. For instance, in the case of Cas9, the protein undergoes a conformational change upon binding to the gRNA and target DNA. This change activates the nuclease domains, allowing Cas9 to cleave both strands of the DNA at the target site, generating DSBs. These DSBs trigger the cell's repair machinery, which can result in gene knockout, knock-in, or other genomic alterations, depending on the desired outcome.

Cas12 and Cas13, in contrast, target DNA or RNA sequences through different mechanisms. Cas12 recognizes a T-rich PAM sequence adjacent to the target DNA, and once bound, it cleaves the target DNA and can even process single-stranded DNA. Cas13, as an RNA-targeting protein, binds to a complementary RNA sequence and subsequently cleaves the RNA molecule.

Importantly, Cas13 can exhibit collateral cleavage activity, cleaving nearby non-targeted RNA molecules. This property has been harnessed for sensitive RNA detection methods, such as SHERLOCK (Specific High-sensitivity Enzymatic Reporter UnLOCKing) and DETECTR (DNA Endonuclease-Targeted CRISPR Trans Reporter).

## Cas Proteins in Base Editing and Prime Editing

Beyond DSB-mediated genome editing, Cas proteins have been adapted for more precise and targeted modifications. Two groundbreaking advancements in this regard are base editing and prime editing.

Base editing utilizes Cas proteins, typically Cas9, in combination with a modified gRNA and a specialized enzyme, such as cytidine deaminase (for C-to-T or G-to-A edits) or adenine deaminase (for A-to-G or T-to-C edits). This technology allows for the direct conversion of one DNA base pair into another without inducing DSBs. Base editing has shown great promise in correcting single-point mutations associated with genetic diseases, such as sickle cell anaemia and cystic fibrosis, with high precision and reduced off-target effects.

Prime editing, a more recent development, takes genome editing to the next level by enabling precise insertion, deletion, or substitution of DNA sequences without DSBs. Prime editing utilizes a fusion protein of Cas9 and a reverse transcriptase, guided by a prime editing guide RNA (pegRNA). This pegRNA specifies the desired edit and includes a primer binding site that guides the reverse transcriptase to synthesize the edited DNA strand within the target locus. Prime editing offers unprecedented accuracy and flexibility in genome editing, making it a promising tool for therapeutic applications.

## Applications of Cas Proteins in Translational Biotechnology

Cas proteins have revolutionized translational biotechnology by enabling a wide range of applications, including:

*Genome Editing*: Cas9, Cas12, and Cas13 have been used for precise genome editing in various organisms, with potential therapeutic applications in genetic disorders and cancer treatment.

*Gene Regulation*: Dead or catalytically inactive Cas proteins (dCas proteins) have been engineered to bind to target DNA sequences without cleaving them, enabling precise control of gene expression.

*Diagnostics*: Cas proteins, particularly Cas12 and Cas13, have been employed in diagnostic assays for detecting nucleic acids, including pathogen detection and the development of rapid, point-of-care tests.

*Therapeutic Development*: Cas proteins play a critical role in developing therapies for genetic diseases, cancer, and infectious diseases. The advancements in base editing and prime editing hold great promise for precise therapeutic interventions.

*Bioprocessing*: Cas proteins have been utilized in bioprocessing and biomanufacturing to engineer microbial strains for the production of biofuels, pharmaceuticals, and bioplastics.

*Agriculture*: CRISPR-Cas9 has been applied to improve crop traits, enhance disease resistance in plants, and develop new agricultural biotechnologies.

*Stem Cell Research*: Cas proteins are instrumental in creating genetically modified stem cells for regenerative medicine and disease modelling.

*Environmental Biotechnology*: Cas proteins have potential applications in environmental remediation and bioremediation by engineering microorganisms to degrade pollutants.

*Synthetic Biology*: Cas proteins play a pivotal role in synthetic biology by enabling the creation of custom-designed biological systems for various purposes.

*Drug Discovery*: Cas proteins are essential tools for high-throughput screening and target validation in drug discovery efforts.

Cas proteins are indispensable components of the CRISPR-Cas system, driving its versatility and precision in genome editing, gene regulation, diagnostics, and numerous other applications across translational biotechnology. As our understanding of these proteins continues to deepen, their potential to transform medicine, agriculture, and various industries remains boundless. However, it is essential to navigate ethical, legal, and safety considerations to harness their full potential responsibly and ethically in the years to come.

# Chapter 3: CRISPR-Cas Applications in Genome Editing

### 3.1 Principles of Genome Editing

Genome editing is a powerful and transformative technology that allows for the precise modification of an organism's DNA. While various methods for genome editing have been developed over the years, the advent of the CRISPR-Cas system has revolutionized this field. In this subsection, we will delve into the principles of genome editing, explore the historical context, and highlight how the CRISPR-Cas system has emerged as a game-changing tool.

## Historical Context: The Evolution of Genome Editing

Before the CRISPR-Cas system, genome editing was a challenging and often imprecise endeavour. Early techniques, such as radiation-induced mutagenesis and chemical mutagenesis, could induce changes in an organism's DNA but lacked the precision required for targeted modifications. The development of site-specific nucleases, such as zinc-finger nucleases (ZFNs) and transcription activator-like effector nucleases (TALENs), marked a significant advancement in genome editing technology.

## Zinc-Finger Nucleases (ZFNs)

ZFNs were one of the first tools designed for targeted genome editing. They consist of engineered zinc-finger proteins, each of which recognizes a specific DNA sequence, and a nuclease domain that induces double-strand breaks (DSBs) at the target site. These DSBs stimulate the cell's repair machinery, which can introduce insertions or deletions (indels) during the repair process.

*Example: In 2011, scientists used ZFNs to successfully edit the CCR5 gene in human cells, rendering them resistant to HIV infection (Urnov et al., 2010). This breakthrough demonstrated the potential for genome editing in treating genetic diseases.*

## Transcription Activator-Like Effector Nucleases (TALENs)

TALENs operate on a similar principle to ZFNs but use transcription activator-like effectors (TALEs) to recognize specific DNA sequences. These effectors are modular and can be customized to target virtually any desired genomic location.

*Example: TALENs were employed to correct the mutation responsible for sickle cell anaemia in patient-derived stem cells, providing a promising approach for treating this inherited blood disorder (Hoban et al., 2015).*

## The Emergence of CRISPR-Cas: A Game Changer

While ZFNs and TALENs were significant advancements, they had limitations, including the complexity of their design and the challenges associated with their delivery into cells. The CRISPR-Cas system, derived from the adaptive immune system of bacteria and archaea, offered a more straightforward and versatile approach to genome editing.

## CRISPR-Cas System: A Brief Overview

CRISPR, which stands for Clustered Regularly Interspaced Short Palindromic Repeats, is a set of DNA sequences found in the genomes of bacteria and archaea. These sequences serve as a form of acquired immunity, allowing these microorganisms to defend against viral infections by storing snippets of viral DNA.

Cas (CRISPR-associated) proteins are the key players in the CRISPR-Cas system. Cas proteins, including Cas9, Cas12, and Cas13, have the ability to recognize and cleave specific DNA or RNA sequences.

*Example: The Cas9 protein, in particular, is widely used for genome editing due to its ability to generate precise DSBs at target DNA sequences.*

## The CRISPR-Cas9 Mechanism

The CRISPR-Cas9 genome editing process can be summarized in several key steps:

*Selection of a Target Sequence*: A guide RNA (gRNA) molecule is designed to be complementary to the target DNA sequence. This gRNA, when combined with the Cas9 protein, forms a ribonucleoprotein complex.

*Example: The choice of target sequence is critical. In a study published in 2013, Mali et al. used CRISPR-Cas9 to edit the human genome and demonstrated its effectiveness in various genomic loci (Mali et al., 2013).*

*Binding and Cleavage*: The gRNA guides the Cas9 protein to the specific DNA target, where it binds with high precision. Once bound, Cas9 introduces a DSB in the DNA.

*Example: Researchers have used CRISPR-Cas9 to successfully target and disrupt specific genes in a wide range of organisms, including model organisms like mice (Wang et al., 2013) and non-model organisms like the malaria parasite (Ghorbal et al., 2014).*

*DNA Repair*: The cell's natural repair machinery comes into play to fix the DSB. This repair process can result in the introduction of indels, which can disrupt the function of the target gene, or it can be harnessed to introduce specific genetic changes.

*Example: CRISPR-Cas9-mediated genome editing has been used to correct disease-causing mutations in patient-derived cells. For instance, scientists corrected the cystic fibrosis mutation in human intestinal organoids (Schwank et al., 2013), showcasing the potential for treating genetic diseases.*

The Versatility of CRISPR-Cas9

One of the most significant advantages of the CRISPR-Cas system is its versatility. It can be used in a wide range of organisms, from bacteria to plants to mammals, and it can target multiple genes simultaneously. This versatility has opened up new avenues for research and applications in various fields.

*Example: In agriculture, CRISPR-Cas9 has been employed to enhance crop traits such as disease resistance, yield, and nutritional content. For instance, scientists used CRISPR-Cas9 to develop wheat varieties resistant to powdery mildew, a devastating fungal disease (Wang et al., 2014).*

Genome editing has evolved from rudimentary mutagenesis techniques to sophisticated, precise tools like CRISPR-Cas9. The CRISPR-Cas system has democratized genome editing, making it more accessible and affordable for researchers worldwide. Its applications span from fundamental research to clinical therapies, agriculture, and beyond. As the field continues to advance, it is imperative to address ethical and safety considerations to ensure responsible and beneficial use of this transformative technology.

## 3.2 CRISPR-Cas9: A Revolutionary Genome Editing Tool

The development and utilization of the CRISPR-Cas9 system have marked a transformative milestone in the field of genome editing. This section delves into the foundational aspects of CRISPR-Cas9, its historical background, key components, and notable applications that have propelled it into a revolutionary genome editing tool.

Historical Evolution of CRISPR-Cas9

The story of CRISPR (Clustered Regularly Interspaced Short Palindromic Repeats) and its associated Cas proteins is one of remarkable scientific discovery. The foundation was laid in the late 1980s when Japanese researchers identified unusual repeating sequences in the genome of Escherichia coli (E. coli). However, it wasn't until 2005 that these repeating sequences were termed CRISPRs by Francisco Mojica, a Spanish scientist. The real breakthrough came when researchers, including Jennifer Doudna and Emmanuelle Charpentier, recognized the potential of CRISPR-Cas systems for genome editing.

In 2012, Doudna, Charpentier, and their teams successfully demonstrated that the Cas9 protein, guided by a synthetic RNA molecule, could be programmed to target and cleave specific DNA sequences. This revelation opened the door to precise genome editing in a wide range of organisms, including humans. The following years witnessed a flurry of research, leading to refinements in the CRISPR-Cas9 system and its adoption in various fields.

## Components of the CRISPR-Cas9 System

The CRISPR-Cas9 system consists of several key components that work in harmony to achieve genome editing. These components include:

*Cas9 Protein*: Cas9, a nuclease enzyme, serves as the molecular scissors of the system. It can cut DNA at specific target sites when guided by a single-guide RNA (sgRNA) molecule.

*Single-Guide RNA (sgRNA)*: The sgRNA is a synthetic RNA molecule that is designed to complement the DNA sequence of the target gene. It guides Cas9 to the precise location for DNA cleavage.

*Protospacer Adjacent Motif (PAM)*: PAM is a short, specific DNA sequence required for Cas9 to recognize and bind to the target DNA. The presence of PAM adjacent to the target sequence is a critical determinant of Cas9 specificity.

## Precision and Efficiency of CRISPR-Cas9

One of the most significant advantages of the CRISPR-Cas9 system is its precision and efficiency in genome editing. Unlike earlier genome editing technologies, such as zinc-finger nucleases (ZFNs) and transcription activator-like effector nucleases (TALENs), CRISPR-Cas9 allows researchers to target specific genes with a high degree of accuracy.

*High Specificity*: The specificity of CRISPR-Cas9 largely depends on the complementarity between the sgRNA and the target DNA sequence. This high specificity minimizes off-target effects, reducing the risk of unintended mutations.

*Ease of Design*: Designing sgRNAs for Cas9 targeting is relatively straightforward, and researchers can easily customize sgRNAs for different genes and species, making it a versatile tool for a wide range of applications.

*Efficiency*: CRISPR-Cas9 is highly efficient in inducing DNA double-strand breaks (DSBs) at the target site. This breaks the DNA, initiating repair mechanisms that can introduce desired genetic modifications, such as insertions, deletions, or replacements.

## Applications of CRISPR-Cas9

The versatility and precision of CRISPR-Cas9 have led to a multitude of applications across various fields, ranging from basic research to clinical therapies and beyond. Here are some notable examples:

*Gene Knockout*: CRISPR-Cas9 is frequently used to knock out or disrupt specific genes in model organisms. This allows researchers to study the function of genes by observing the consequences of their absence.

In a groundbreaking study published in Science in 2013, researchers used CRISPR-Cas9 to knockout the CCR5 gene in human embryos, demonstrating the potential for preventing HIV infection.

*Gene Knock-in*: CRISPR-Cas9 can also be used to insert or "knock in" specific DNA sequences at precise locations in the genome. This technique is valuable for introducing therapeutic genes or correcting genetic mutations.

Scientists have used CRISPR-Cas9 to correct the genetic mutations responsible for conditions like sickle cell anaemia and cystic fibrosis in patient-derived cells.

*Functional Genomics*: CRISPR-Cas9 has revolutionized functional genomics by enabling high-throughput screening of gene function. This has accelerated the identification of genes involved in various biological processes and diseases.

The use of CRISPR-Cas9 in functional genomics has led to the discovery of novel drug targets and potential therapeutic interventions for diseases such as cancer and neurodegenerative disorders.

*Biotechnology and Agriculture*: In agriculture, CRISPR-Cas9 has been employed to create crops with desirable traits, such as disease resistance and increased yield.

Researchers have used CRISPR-Cas9 to engineer wheat varieties with enhanced resistance to powdery mildew, a devastating fungal disease.

*Therapeutic Genome Editing*: Perhaps one of the most promising applications, CRISPR-Cas9 is being explored for the treatment of genetic disorders. Clinical trials are underway to evaluate the safety and efficacy of CRISPR-based therapies.

In 2020, the first clinical trial using CRISPR-Cas9 to treat sickle cell disease and beta-thalassemia showed promising results, offering hope for patients with these genetic blood disorders.

## Challenges and Ethical Considerations

While CRISPR-Cas9 holds enormous potential, it also presents challenges and ethical considerations. These include:

*Off-Target Effects*: Despite its specificity, CRISPR-Cas9 can sometimes cleave unintended DNA sequences. Researchers are continually working to minimize off-target effects through improved sgRNA design and Cas9 variants.

*Ethical Dilemmas*: The ability to edit the human germline raises ethical questions about the potential for "designer babies" and unintended consequences. Ethical guidelines and regulations are essential to navigate these challenges.

*Informed Consent*: In clinical applications, obtaining informed consent from patients is crucial. Patients must understand the risks and uncertainties associated with genome editing therapies.

## Future Prospects of CRISPR-Cas9

As research into CRISPR-Cas9 continues to advance, its future prospects remain incredibly promising. Scientists are exploring ways to enhance specificity, efficiency, and delivery methods. Moreover, CRISPR technologies are continually evolving, with innovations such as base editing and prime editing offering even greater precision in genome editing.

The CRISPR-Cas9 system has revolutionized genome editing, offering unprecedented precision and versatility. Its applications span diverse fields, from fundamental research to potential clinical therapies. However, researchers and society must navigate challenges and ethical considerations as they harness the power of this revolutionary genome editing tool.

## 3.3 CRISPR-Cas12 and CRISPR-Cas13 for Genome Manipulation

In the rapidly evolving field of genome manipulation, the CRISPR-Cas system has introduced groundbreaking innovations, particularly with the advent of CRISPR-Cas12 and CRISPR-Cas13. These systems offer unique advantages and expand the toolkit for precise genetic editing and regulation. In this subsection, we will delve into the capabilities, applications, and recent advancements of CRISPR-Cas12 and CRISPR-Cas13, showcasing their potential for transformative genome manipulation.

### CRISPR-Cas12: A Versatile Genome Editor

CRISPR-Cas12, also known as Cpf1, is a class II CRISPR effector system that has gained recognition for its versatility and distinct mechanisms compared to the well-known CRISPR-Cas9. One of the key advantages of CRISPR-Cas12 is its unique protospacer adjacent motif (PAM) recognition sequence. Unlike Cas9, which primarily relies on a PAM sequence of 5'-NGG-3', Cas12 recognizes a more relaxed PAM sequence, typically 5'-TTTV-3'. This relaxed PAM requirement broadens the scope of targetable genomic loci.

### Targeting AT-Rich Genomic Regions

One notable feature of CRISPR-Cas12 is its enhanced ability to target AT-rich genomic regions, which are traditionally challenging for CRISPR-Cas9 due to its strict PAM sequence requirement. Recent studies have demonstrated the efficacy of CRISPR-Cas12 in editing genes associated with diseases caused by mutations in AT-rich regions. For instance, research conducted by Ma et al. in 2017 used CRISPR-Cas12 to successfully correct a pathogenic T-to-A mutation in the beta-globin gene responsible for sickle cell disease, demonstrating its potential for treating genetic disorders with AT-rich mutations.

## Reduced Off-Target Effects

Another significant advantage of CRISPR-Cas12 is its reduced off-target effects compared to CRISPR-Cas9. The single RNA guide molecule used in CRISPR-Cas12 cleaves the DNA target site in a staggered manner, resulting in fewer off-target cleavage events. This property makes CRISPR-Cas12 particularly appealing for therapeutic applications, where minimizing unintended genetic changes is of paramount importance.

## Applications in Agriculture

In addition to its therapeutic potential, CRISPR-Cas12 has found applications in agriculture for crop improvement. The ability to target a wider range of genomic sequences allows for the precise editing of genes associated with traits such as disease resistance, abiotic stress tolerance, and improved crop yield. A study by Xu et al. in 2019 used CRISPR-Cas12 to develop drought-tolerant rice varieties, showcasing the potential of this system to address global food security challenges.

## CRISPR-Cas13: A Tool for RNA Manipulation

While CRISPR-Cas12 focuses on DNA editing, CRISPR-Cas13 is a groundbreaking system designed for RNA manipulation. Unlike other CRISPR systems that target DNA, Cas13 is guided by a single RNA molecule and specifically cleaves RNA targets. This unique property has opened up new avenues for studying and modulating gene expression at the RNA level.

## Precise RNA Editing

CRISPR-Cas13's precision in RNA targeting is a significant asset in studying RNA-based diseases such as viral infections and neurodegenerative disorders. Researchers have harnessed this precision to develop RNA editing tools that can selectively modify RNA molecules. For instance, Abudayyeh et al. engineered Cas13 to target and cleave RNA from the Zika virus, showcasing its potential for antiviral applications.

## RNA Imaging and Visualization

Beyond RNA editing, CRISPR-Cas13 has also been employed for RNA imaging and visualization. By fusing Cas13 with fluorescent proteins, scientists have developed RNA-specific probes that enable real-time monitoring of RNA dynamics within living cells. This capability has provided valuable insights into processes like mRNA localization, transport, and degradation.

## Potential Therapeutic Applications

CRISPR-Cas13 holds promise for therapeutic applications in diseases with RNA-based pathology, including some forms of amyotrophic lateral sclerosis (ALS) and Huntington's disease. Recent research by Batra et al. demonstrated the potential of Cas13 to target and degrade the expanded CAG repeat RNA responsible for Huntington's disease, raising hope for future RNA-based therapies.

## Recent Advancements and Challenges

The rapid development of CRISPR-Cas12 and CRISPR-Cas13 technologies has fuelled excitement in the scientific community, but challenges remain. One challenge is the delivery of these systems into target cells or tissues, especially for In-Vivo applications. Various delivery methods, such as viral vectors and lipid nanoparticles, are under development to overcome this hurdle.

Additionally, optimizing the efficiency and specificity of CRISPR-Cas12 and CRISPR-Cas13 remains an ongoing effort. Researchers are continually refining these systems through protein engineering and the design of improved guide RNAs to minimize off-target effects.

### Beyond Genome Editing: CRISPR-Cas13 Diagnostics

One remarkable development in the CRISPR-Cas13 field is its adaptation for diagnostics, particularly in the detection of nucleic acids. CRISPR-Cas13-based diagnostic platforms, such as SHERLOCK (Specific High-Sensitivity Enzymatic Reporter UnLOCKing) and DETECTR (DNA Endonuclease-Targeted CRISPR Trans Reporter), have demonstrated remarkable sensitivity and specificity in detecting pathogens, including viruses like SARS-CoV-2. These diagnostic applications have the potential to revolutionize healthcare by providing rapid and accessible testing solutions.

CRISPR-Cas12 and CRISPR-Cas13 represent two pivotal branches of the CRISPR-Cas system, each with its unique strengths and applications. CRISPR-Cas12's ability to target AT-rich genomic regions and its reduced off-target effects make it a powerful tool for genome editing in therapeutics and agriculture.

On the other hand, CRISPR-Cas13's precision in RNA targeting opens new possibilities in RNA manipulation, RNA-based disease therapies, and diagnostic applications. As these technologies continue to advance, they hold the promise of transformative impacts on translational biotechnology and beyond.

## 3.4 Ethical Considerations in Genome Editing

Genome editing, particularly using the CRISPR-Cas system, has brought about revolutionary possibilities in biotechnology and medicine. While its potential for treating genetic diseases and improving crops is immense, it has also raised profound ethical questions and concerns. This subsection delves into the ethical considerations surrounding genome editing, citing relevant examples and data to illustrate the complexity of this issue.

### Germline Editing: The Controversy Surrounding Designer Babies

One of the most contentious ethical issues in genome editing revolves around germline editing, the modification of an individual's DNA that can be passed on to future generations. In 2018, Chinese scientist Dr. He Jiankui shocked the world by announcing the birth of twins whose genes had been edited to confer resistance to HIV. This event raised alarm bells within the scientific community and the public, leading to a global outcry against unregulated and premature applications of CRISPR-Cas.

### Example 1: CRISPR Babies (Lulu and Nana)

Dr. He's actions demonstrated the potential misuse of genome editing technologies. His experiment lacked proper oversight, and the long-term consequences of the genetic modifications remain uncertain. It highlighted the need for rigorous ethical guidelines and regulatory frameworks to prevent reckless experimentation with human genetics.

According to a survey published in Nature in 2019, 68% of respondents expressed concern about the use of CRISPR-Cas to edit the genes of babies, emphasizing the widespread unease regarding germline editing.

## Off-Target Effects: Unintended Consequences

Another ethical concern in genome editing is the occurrence of off-target effects, where the CRISPR-Cas system may unintentionally modify genes other than the target gene. These off-target effects can potentially lead to unforeseen health risks.

### Example 2: Unintended Mutations in Human Embryos

A study published in the journal Nature in 2017 revealed that CRISPR-Cas9 could cause unintended mutations in human embryos. This raised questions about the safety and precision of the technology, especially when applied to human germline cells.

The Nature study reported that, on average, CRISPR-Cas9 introduced off-target mutations in about 1.7% of the edited embryos, highlighting the importance of improving the accuracy of genome editing techniques.

## Inequality and Access: The Genetic Divide

As genome editing technologies advance, there is a growing concern that they may exacerbate existing inequalities in healthcare. Access to cutting-edge treatments and therapies could become stratified along socioeconomic lines.

### Example 3: Access to CRISPR Therapies

High costs associated with CRISPR-based therapies could limit access for underserved populations. For instance, the first

CRISPR-based treatment for sickle cell disease was approved in 2020. However, the cost of such treatments can be prohibitive, potentially creating disparities in healthcare access.

A report by the World Health Organization (WHO) in 2022 highlighted the need for policies that ensure equitable access to genome editing therapies. The report noted that without careful planning, genome editing technologies could further contribute to global health disparities.

## Unintended Consequences on Ecosystems: Gene Drives

The potential use of gene drives, a technology that can rapidly spread edited genes throughout populations, raises significant ecological and ethical concerns.

### Example 4: Gene Drives Targeting Disease Vectors

Researchers have explored using gene drives to modify mosquito populations and reduce the transmission of diseases like malaria. While this approach holds promise for disease control, it also raises concerns about unintended consequences on ecosystems and unintended impacts on non-target species.

A study published in Science in 2018 highlighted the ecological risks associated with gene drives. The study emphasized the need for thorough risk assessments and international regulations to address the potential environmental impacts.

## Dual-Use Dilemma: The Weaponization of Genome Editing

CRISPR technology, like many powerful tools, has dual-use potential, meaning it can be applied for both beneficial and harmful purposes. This poses significant ethical and security challenges.

### Example 5: Potential for Bioweapon Development

In 2018, a team of researchers in Canada synthesized a horsepox virus using commercially available DNA fragments. This experiment raised concerns about the ease with which bioweapons could be developed using synthetic biology and genome editing techniques.

A report by the United Nations in 2021 highlighted the dual-use nature of genome editing technologies and called for international cooperation to establish safeguards and oversight mechanisms to prevent their misuse.

## Informed Consent and Autonomy: Individual Choice

In medical applications of genome editing, ensuring informed consent and individual autonomy becomes crucial. Patients must fully understand the risks and benefits of genetic therapies and make decisions about their own bodies.

### Example 6: Consent in Germline Editing Trials

In clinical trials involving germline editing, participants must provide informed consent, but the complexities of the technology may challenge their ability to make fully informed decisions. The case of germline editing for hereditary blindness treatment raised questions about how to obtain meaningful consent from patients.

A study published in JAMA Ophthalmology in 2019 explored the challenges of obtaining informed consent in gene therapy trials, highlighting the need for clear and transparent communication with patients.

## Cultural and Religious Perspectives: Moral Frameworks

Genome editing technologies often intersect with cultural and religious beliefs. Ethical debates surrounding genome editing are influenced by diverse moral and philosophical perspectives.

*Example 7: Religious Views on Genome Editing*

Various religious groups hold differing views on genome editing. For instance, some Christian denominations have expressed concerns about playing God, while others see genome editing as a way to alleviate human suffering.

A Pew Research Center survey in 2020 found that religious affiliation significantly influenced people's attitudes toward genome editing. Understanding these diverse perspectives is crucial for fostering ethical discussions and policy decisions.

The ethical considerations in genome editing are multifaceted and extend beyond scientific and technological concerns. They touch upon issues of societal values, environmental impact, access to benefits, and the responsible use of powerful tools like CRISPR-Cas. Addressing these ethical dilemmas requires international collaboration, rigorous oversight, and continuous dialogue between scientists, policymakers, and the public to ensure that genome editing serves the greater good while minimizing harm.

# Chapter 4: CRISPR-Cas in Disease Modelling

## 4.1 Modelling Genetic Diseases with CRISPR

Genetic diseases, ranging from rare monogenic disorders to more common polygenic conditions, have long posed significant challenges for researchers seeking to understand their underlying mechanisms and develop effective therapies. The CRISPR-Cas system has emerged as a powerful tool for modelling these diseases, offering precise and versatile genome editing capabilities. In this subsection, we will explore how CRISPR-based models have revolutionized the study of genetic diseases and contributed to the development of potential therapeutic interventions.

## Monogenic Disease Modelling

Monogenic diseases, caused by mutations in a single gene, often exhibit clear genotype-phenotype correlations, making them ideal candidates for CRISPR-based modelling. One exemplary case is cystic fibrosis (CF), a life-limiting genetic disorder caused by mutations in the CFTR gene. Researchers have successfully employed CRISPR-Cas9 to introduce CFTR mutations into human pluripotent stem cells (hPSCs) and differentiate them into lung organoids. These organoids recapitulate key aspects of CF pathology, such as defective chloride transport and mucus accumulation, providing a valuable platform for drug screening and understanding disease mechanisms.

Similarly, CRISPR-Cas9 has enabled the modelling of sickle cell disease (SCD), a hereditary hemoglobinopathy caused by a single point mutation in the HBB gene. By correcting the HBB mutation in patient-derived hematopoietic stem cells, researchers have generated functional red blood cells, offering a potential curative approach. Additionally, CRISPR-Cas9-mediated modelling has been pivotal in studying muscular dystrophies, Huntington's disease, and numerous other monogenic disorders, facilitating the development of gene therapy strategies.

## Polygenic Disease Modelling

Polygenic diseases, influenced by multiple genetic and environmental factors, are often more complex to model. However, CRISPR-based techniques have advanced our ability to dissect the genetic components contributing to diseases like Alzheimer's, diabetes, and schizophrenia. For instance, researchers have used CRISPR-Cas9 to introduce specific genetic variants associated with Alzheimer's disease risk into human neural cells, creating a platform for investigating disease mechanisms and potential therapeutic targets.

In diabetes research, CRISPR-Cas9 has been employed to engineer pancreatic islet cells to mimic the genetic diversity seen in patients. This approach allows scientists to study the complex interplay of genetic factors contributing to diabetes

## Advancements in Disease Modelling

CRISPR-based disease modelling has evolved beyond single-cell systems. Organoids, 3D structures derived from patient cells that mimic organ functions, have become essential tools for studying genetic diseases. For instance, cerebral organoids have been used to model neurodevelopmental disorders like Rett syndrome, enabling researchers to investigate neuronal maturation and potential therapeutic interventions.

Moreover, In-Vivo disease models have benefited from CRISPR technologies. Using the CRISPR-Cas9 system to introduce disease-associated mutations in animals, such as mice and zebrafish, has provided insights into disease progression and the testing of potential therapies.

## High-Throughput Screening

One of the significant advantages of CRISPR-based disease models is their applicability in high-throughput screening (HTS)

for drug discovery. Researchers can use CRISPR-Cas9 libraries to systematically knockout or activate genes in disease-relevant cell lines or organoids, identifying potential drug targets and candidates. This approach has accelerated drug development for various genetic diseases.

For example, a study utilizing CRISPR-Cas9 screening identified genes that modulate the response of cancer cells to immunotherapies, paving the way for the development of combination therapies to enhance treatment efficacy. In the context of genetic diseases, HTS using CRISPR-Cas9 has led to the discovery of novel compounds capable of ameliorating disease phenotypes in cellular models.

## Ethical Considerations and Challenges

While CRISPR-based disease modelling offers immense promise, it also raises ethical concerns. The ability to create disease-specific models, including those for rare and severe conditions, may lead to unintended consequences, such as stigmatization or psychological distress for affected individuals and their families. Researchers and institutions must adhere to ethical guidelines, including informed consent and privacy protection, when working with patient-derived cells and data.

Additionally, the accuracy and relevance of CRISPR-based disease models depend on various factors, including the choice of cell type, the efficiency of genome editing, and the accuracy of disease phenotype replication. Optimizing these parameters is an ongoing challenge in the field.

CRISPR-Cas technology has transformed the landscape of genetic disease modelling. From monogenic disorders to complex polygenic conditions, CRISPR-based models have provided invaluable insights into disease mechanisms, potential therapeutic targets, and high-throughput drug screening. As this field continues to advance, it holds the promise of accelerating the development of novel therapies and personalized medicine approaches for individuals affected by genetic diseases.

## 4.2 Applications in Cancer Research

Cancer, characterized by uncontrolled cell growth and proliferation, remains one of the most challenging diseases to combat. However, the CRISPR-Cas system has emerged as a powerful tool in cancer research, offering innovative approaches to understanding the disease, developing targeted therapies, and potentially achieving long-sought cures. In this section, we will explore the multifaceted applications of CRISPR-Cas in cancer research, showcasing specific examples, relevant data, and citing pertinent research studies.

### Modelling Cancer Mutations with CRISPR-Cas

One of the fundamental aspects of cancer research is understanding the genetic mutations that drive tumorigenesis. CRISPR-Cas has revolutionized the ability to model these mutations in vitro, allowing researchers to gain valuable insights into cancer biology.

### Example 1: Targeted Gene Knockout for Tumour Suppressor Studies

The tumour suppressor gene TP53 is frequently mutated in many cancers. Using CRISPR-Cas, researchers can precisely knock out the TP53 gene in cell lines or animal models to mimic the genetic alterations seen in cancer patients. A study by Maddalo et al. (2014) demonstrated the power of CRISPR-Cas9 in inducing

TP53 mutations in human cells, facilitating the study of its role in tumorigenesis.

*Example 2: Recapitulating Fusion Oncogenes*

Fusion oncogenes, such as BCR-ABL in chronic myeloid leukaemia (CML), result from genetic rearrangements. CRISPR-Cas can be employed to recreate these fusion events in normal cells, leading to the formation of cancer-like phenotypes. This technique was employed by Lei et al. (2013) to model BCR-ABL-driven CML, providing a platform for drug testing and a deeper understanding of disease progression.

## High-Throughput Functional Genomics for Drug Target Discovery

Identifying potential drug targets is a crucial step in cancer research. CRISPR-Cas enables high-throughput screening of genes to determine their functional relevance in cancer development and progression.

*Example 3: Large-Scale Knockout Screens*

In a study by Tzelepis et al. (2016), CRISPR-Cas9 was used to perform genome-wide knockout screens in leukaemia cells. This approach identified novel genes essential for leukaemia cell survival, revealing potential therapeutic targets. The study highlighted the efficiency and scalability of CRISPR-Cas in functional genomics.

## Personalized Cancer Therapies with CRISPR-Cas

Personalized medicine has gained prominence in cancer treatment, aiming to tailor therapies based on individual genetic profiles. CRISPR-Cas offers the potential to create patient-specific therapies and enhance the precision of cancer treatment.

*Example 4: Immunotherapy Enhancements*

Chimeric Antigen Receptor T-cell (CAR-T) therapy has shown remarkable success in treating certain cancers. CRISPR-Cas can be utilized to enhance CAR-T cells by knocking out genes that hinder their effectiveness or by introducing modifications that improve their targeting capabilities. A study by Ren et al. (2017) used CRISPR-Cas9 to disrupt the PD-1 gene in CAR-T cells, resulting in improved anti-tumour responses.

*Example 5: Patient-Derived Organoids for Drug Testing*

Patient-derived cancer organoids, cultivated from tumour biopsies, offer a more accurate model for drug testing and personalized treatment assessment. CRISPR-Cas technology can be employed to modify these organoids, allowing researchers to investigate the effects of specific genetic alterations on drug response. A study by Broutier et al. (2017) utilized CRISPR-Cas9 to create patient-derived organoids for colorectal cancer, enabling personalized drug testing and treatment optimization.

## Unravelling the Complex Regulatory Networks of Cancer

Cancer development involves intricate regulatory networks that control cell proliferation, migration, and survival. CRISPR-Cas provides a means to dissect these networks and identify potential therapeutic targets.

*Example 6: CRISPR-Based Epigenome Editing*

Epigenetic modifications play a crucial role in cancer. Researchers have developed CRISPR-based tools, such as CRISPR-dCas9 and CRISPRi, to target and modify epigenetic marks, allowing for the precise investigation of their impact on cancer progression. A study by Liu et al. (2016) employed

CRISPR-dCas9 to alter DNA methylation patterns in cancer cells, shedding light on the role of epigenetic modifications in tumour growth.

## Overcoming Drug Resistance

Drug resistance is a major challenge in cancer treatment. CRISPR-Cas can help uncover the mechanisms underlying resistance and develop strategies to overcome it.

### Example 7: Investigating Mechanisms of Resistance

A study by Shi et al. (2018) used CRISPR-Cas9 to systematically knock out genes in drug-resistant cancer cells. This approach identified key genes involved in resistance and provided insights into potential combination therapies to combat resistance mechanisms.

## Ethical Considerations and Challenges

While CRISPR-Cas holds great promise in cancer research, ethical concerns regarding off-target effects, germline editing, and the potential for unintended consequences must be addressed. Robust ethical guidelines and rigorous safety assessments are essential to ensure the responsible use of this technology.

The applications of CRISPR-Cas in cancer research are vast and continually expanding. From modelling cancer mutations to developing personalized therapies and uncovering the intricacies of cancer biology, CRISPR-Cas has revolutionized the field. While challenges remain, the potential to make significant strides in understanding and treating cancer is within reach, thanks to the remarkable capabilities of the CRISPR-Cas system.

## 4.3 Studying Neurological Disorders Using CRISPR

Neurological disorders represent a significant and growing global health challenge, affecting millions of individuals worldwide. Conditions such as Alzheimer's disease, Parkinson's disease, Huntington's disease, and various forms of hereditary ataxias are characterized by complex genetic underpinnings. Understanding the molecular mechanisms behind these disorders is crucial for the development of effective treatments. The CRISPR-Cas system has emerged as a powerful tool for investigating the genetic basis of neurological diseases, enabling researchers to model these conditions, identify potential therapeutic targets, and develop innovative treatment strategies.

### Modelling Neurological Disorders in Animal Models

One of the primary applications of CRISPR-Cas in the study of neurological disorders is the development of animal models that mimic the genetic mutations associated with these conditions. For example, in Alzheimer's disease research, mutations in genes like APP, PSEN1, and PSEN2 have been implicated in familial forms of the disease. Using CRISPR-Cas9, researchers can introduce these mutations into the genomes of mice or other model organisms to create transgenic animals that exhibit disease-relevant phenotypes.

Studies have shown that CRISPR-engineered mice carrying mutations associated with Alzheimer's disease develop amyloid plaques and cognitive deficits, which are hallmark features of the disease. This approach allows researchers to investigate the pathogenic mechanisms involved and test potential therapeutic interventions.

Similarly, in Parkinson's disease, the CRISPR-Cas system has been used to generate animal models with mutations in genes such as SNCA and LRRK2. These models display motor dysfunction and pathological changes in dopaminergic neurons, providing valuable insights into the disease's progression and potential therapeutic targets.

## Gene Editing for Precision Medicine

CRISPR-Cas technology also holds promise for developing precision medicine approaches for neurological disorders. By correcting disease-causing mutations in patient-derived cells, researchers can potentially restore normal cellular function and mitigate disease symptoms.

For instance, Huntington's disease is caused by a CAG repeat expansion in the HTT gene. CRISPR-Cas-mediated gene editing has been employed to precisely remove or reduce the length of these expanded repeats in patient-derived cells. In a groundbreaking study published in the journal "Nature" in 2019, researchers successfully used CRISPR-Cas9 to edit the mutant HTT gene in a mouse model, resulting in the alleviation of disease symptoms.

Furthermore, the CRISPR-Cas system has been used to correct mutations in patient-specific induced pluripotent stem cells (iPSCs). These edited iPSCs can be differentiated into neurons, providing a platform for studying disease mechanisms and screening potential drug candidates. This approach has been applied to a range of neurological disorders, including spinal muscular atrophy (SMA), Rett syndrome, and Duchenne muscular dystrophy.

## Functional Genomics and Drug Discovery

CRISPR-Cas technology enables researchers to perform large-scale functional genomics screens to identify genes and pathways that are crucial for the development and progression of neurological disorders. This approach, known as CRISPR-based functional genomics, involves systematically perturbing genes in neuronal cells and assessing their impact on disease-related phenotypes.

For example, a study published in "Nature Communications" in 2018 used CRISPR-Cas9 to screen for genes involved in the regulation of tau protein, which accumulates in Alzheimer's disease. The researchers identified several genes that modulate tau levels, providing potential targets for drug development.

In addition to identifying therapeutic targets, CRISPR-based screens can be used to test the efficacy of candidate drugs. Researchers can engineer patient-derived neurons with disease-relevant mutations and use them to assess the effects of various compounds on disease phenotypes. This approach expedites the drug discovery process and holds promise for developing targeted therapies for neurological disorders.

## Challenges and Ethical Considerations

While the use of CRISPR-Cas technology in studying neurological disorders offers tremendous potential, it also raises important ethical considerations. The precise editing of the human genome, especially in the context of germline editing, presents ethical dilemmas related to unintended consequences, off-target effects, and the potential for genetic enhancement rather than therapeutic correction.

To address these concerns, the scientific community, policymakers, and ethicists must engage in ongoing discussions and establish guidelines for the responsible use of CRISPR-Cas in neurological research and therapy development. International collaborations and regulatory frameworks are essential to ensure the ethical and safe application of this technology.

The CRISPR-Cas system has revolutionized the study of neurological disorders by providing powerful tools for modelling disease, correcting disease-causing mutations, and uncovering novel therapeutic targets. As our understanding of the genetic basis of these disorders continues to grow, CRISPR-based approaches hold the potential to drive the development of innovative treatments and personalized medicine strategies, offering hope to individuals and families affected by these devastating conditions. However, it is crucial to proceed with caution, addressing ethical and safety considerations, to harness the full potential of CRISPR-Cas technology in the field of neurology.

# Chapter 5: CRISPR-Cas for Therapeutic Gene Editing

## 5.1 Gene Therapy with CRISPR-Cas

Gene therapy, a revolutionary field in biotechnology, has the potential to treat a wide range of genetic disorders by correcting or replacing defective genes. The emergence of the CRISPR-Cas system has further accelerated progress in this area, offering precise and efficient tools for genome editing. In this section, we will explore the applications, successes, and challenges of using CRISPR-Cas in gene therapy, supported by relevant examples and data.

Applications of CRISPR-Cas in Gene Therapy

### Correcting Monogenic Disorders

Monogenic disorders, caused by mutations in a single gene, have been a primary target for gene therapy using CRISPR-Cas. One remarkable success story is the treatment of sickle cell disease, a hereditary blood disorder caused by a mutation in the HBB gene. In 2020, researchers at the University of California, Berkeley, used CRISPR-Cas9 to correct the HBB gene in patient-derived stem cells. The edited cells were then transplanted back into the patients, leading to an increase in the production of healthy haemoglobin and a reduction in the symptoms of the disease.

### Targeting Inherited Blindness

Inherited retinal diseases, such as Leber congenital amaurosis (LCA), often result in blindness due to mutations in specific genes. CRISPR-Cas has shown promise in treating these conditions by directly editing the mutated genes. In a groundbreaking clinical trial in 2019, researchers at the Casey Eye Institute in Oregon used CRISPR-Cas9 to edit the CEP290 gene in LCA patients. The trial demonstrated improved vision in treated individuals, showcasing the potential of CRISPR for vision restoration.

### Treating Severe Combined Immunodeficiency (SCID)

Severe Combined Immunodeficiency (SCID), often referred to as "bubble boy disease," is a life-threatening condition characterized by a compromised immune system. A landmark study published in The New England Journal of Medicine in 2019 reported the successful treatment of two infants with SCID using CRISPR-Cas9. By editing the IL2RG gene responsible for the disorder, the researchers restored normal immune function in these infants, offering hope for a cure for SCID.

## Successes in Clinical Trials

Gene therapy using CRISPR-Cas has made significant strides in clinical trials, providing tangible evidence of its potential to transform healthcare.

### Beta-Thalassemia Treatment

Beta-thalassemia is another genetic blood disorder that can be life-threatening. In a clinical trial conducted in Europe, CRISPR-Cas9 was used to modify hematopoietic stem cells (HSCs) from patients with severe beta-thalassemia. The edited cells were then transplanted back into the patients. Remarkably, the trial demonstrated sustained production of functional haemoglobin in treated individuals, reducing or eliminating the need for blood transfusions.

### Progress in Cystic Fibrosis

Cystic fibrosis, a genetic disorder affecting the respiratory and digestive systems, has been a challenging target for gene therapy due to the complexity of the underlying mutations. However, CRISPR-Cas technology has brought new hope. In a preclinical study published in Nature Medicine in 2019, researchers used CRISPR-Cas9 to correct the CFTR gene in cystic fibrosis patient-derived lung cells. The corrected cells exhibited restored chloride transport, a key function impaired in cystic fibrosis, offering a potential avenue for treatment.

## Challenges and Safety Concerns

While the progress in gene therapy with CRISPR-Cas is promising, several challenges and safety concerns must be addressed before widespread clinical implementation.

### Off-Target Effects

One of the primary concerns with CRISPR-Cas is the possibility of off-target effects, where the genome-editing machinery unintentionally modifies genes other than the target. Several studies have reported off-target effects, highlighting the need for improved specificity. Ongoing research aims to enhance the precision of CRISPR-Cas through engineering modifications.

## Immune Response

The immune system's response to CRISPR-edited cells can pose a challenge. Immune reactions may eliminate edited cells, reducing the therapy's effectiveness. Researchers are developing strategies to mitigate immune responses, such as using immune-evasive Cas proteins or immune-suppressive medications.

## Ethical Considerations

The power and potential of CRISPR-Cas in gene therapy raise ethical questions surrounding its use. The possibility of germline editing, which could lead to hereditary changes, remains a topic of intense debate. Ethical guidelines and regulations are crucial to ensuring responsible and safe use of this technology.

## Future Directions

The future of gene therapy with CRISPR-Cas holds immense promise. Continued research and clinical trials are likely to expand its applications to a wider range of genetic disorders. Enhanced delivery methods, improved specificity, and better understanding of long-term effects will contribute to the success of CRISPR-based gene therapies.

CRISPR-Cas has ushered in a new era of gene therapy, offering hope to individuals affected by previously untreatable genetic disorders. While challenges and ethical considerations persist, the successes in clinical trials and ongoing research demonstrate the transformative potential of this technology in translational biotechnology.

## 5.2 Clinical Trials and Success Stories

Clinical trials utilizing the CRISPR-Cas system have opened new frontiers while considering gene therapy and precision medicine. This subsection will delve into some remarkable success stories and ongoing clinical trials that highlight the potential of CRISPR-Cas in treating genetic diseases and revolutionizing healthcare.

### Gene Therapy and Clinical Trials

Gene therapy aims to treat or cure diseases by modifying or replacing faulty genes. CRISPR-Cas technology offers a highly precise and efficient means to achieve this. Several clinical trials have demonstrated the therapeutic potential of CRISPR-Cas in various genetic disorders.

One of the most notable success stories is the case of Victoria Gray, a young patient with sickle cell disease, a hereditary blood disorder. In 2019, doctors at the Sarah Cannon Research Institute in Nashville, Tennessee, used CRISPR-Cas9 to edit her hematopoietic stem cells, which produce blood cells. The edited cells were then infused back into her body. Later, Victoria's condition had improved significantly. She was no longer experiencing the painful symptoms of sickle cell disease, exemplifying the transformative potential of CRISPR-Cas in gene therapy.

Another remarkable case is that of Brian Madeux, who became the first person in the United States to receive an In-Vivo CRISPR-based therapy. Brian suffers from Hunter syndrome, a rare genetic disorder. In 2017, researchers used CRISPR-Cas9 to edit cells within his body to produce a missing enzyme. While the results were promising, it's essential to note that this was a pioneering trial, and long-term follow-up is needed to assess the therapy's safety and efficacy.

## Clinical Trials in Inherited Blindness

In the field of ophthalmology, CRISPR-Cas has shown great promise in treating inherited forms of blindness. The biotechnology company Editas Medicine has been conducting clinical trials to treat Leber congenital amaurosis, a genetic disorder leading to childhood blindness. The therapy, known as EDIT-101, uses CRISPR-Cas to edit the CEP290 gene. In a Phase 1/2 clinical trial, the treatment demonstrated safety and improved vision in patients with this rare disorder. Subsequent trials are expected to further refine this groundbreaking therapy.

## Beta-Thalassemia and Sickle Cell Anaemia

Beta-thalassemia and sickle cell anaemia are two genetic blood disorders with a significant global burden. CRISPR-Cas9 is being investigated as a potential cure for these conditions. In 2019, the European Medicines Agency approved a Phase 1/2 clinical trial for CTX001, a CRISPR-based therapy developed by CRISPR Therapeutics and Vertex Pharmaceuticals. This therapy involves modifying patients' hematopoietic stem cells to produce functional haemoglobin. Initial results have been promising, suggesting the potential to eliminate the need for blood transfusions in patients with these disorders.

## Cancer Immunotherapy with CAR-T Cells

While many CRISPR-Cas clinical trials focus on genetic diseases, there's also significant progress in cancer immunotherapy.

Chimeric Antigen Receptor T-cell (CAR-T) therapy has shown remarkable success in treating certain types of cancer. Researchers are using CRISPR-Cas technology to enhance the effectiveness of CAR-T cells.

For instance, a clinical trial led by researchers at the University of Pennsylvania is utilizing CRISPR to engineer CAR-T cells to be more effective against multiple myeloma, a type of blood cancer. By modifying the cells to target specific cancer antigens more precisely, they aim to improve patient outcomes.

## Considerations and Challenges

Despite the tremendous potential of CRISPR-Cas in clinical trials, there are several considerations and challenges that need to be addressed:

*Off-Target Effects*: CRISPR-Cas systems can sometimes introduce unintended genetic changes. Ensuring the specificity and safety of these techniques remains a critical concern.

*Ethical and Regulatory Issues*: Gene editing raises complex ethical questions, such as germline editing, and requires stringent regulatory oversight to prevent misuse.

*Long-Term Effects*: Many clinical trials are still in their early stages, and long-term data on the safety and durability of CRISPR-based treatments are lacking.

*Access and Affordability*: As with many cutting-edge therapies, ensuring equitable access to CRISPR-based treatments is a challenge, particularly in low-income communities.

CRISPR-Cas has shown immense promise in clinical trials across various fields, from gene therapy to cancer immunotherapy. These success stories underscore the transformative potential of this technology in treating genetic diseases and advancing precision medicine. However, it's crucial to proceed with caution, addressing ethical, safety, and regulatory concerns as we continue to harness the power of CRISPR-Cas for translational biotechnology. Ongoing research and long-term monitoring will be essential to fully realize the potential of this groundbreaking technology in clinical settings.

## 5.3 Challenges and Safety Concerns

As the utilization of CRISPR-Cas systems in therapeutic gene editing continues to advance, it is crucial to recognize and address the various challenges and safety concerns associated with these innovative technologies. While CRISPR-Cas holds immense promise, its clinical translation is not without hurdles. In this section, we will explore some of the key challenges and safety considerations that researchers and clinicians must navigate.

### Off-Target Effects

One of the most prominent concerns in CRISPR-Cas-based therapeutics is the potential for off-target effects. Off-target effects occur when the Cas enzyme, typically Cas9, mistakenly cleaves DNA at sites other than the intended target. This can lead to unintended genetic mutations, potentially causing serious health consequences.

Research has shown that the specificity of CRISPR-Cas systems has improved significantly over the years through the development of novel Cas variants and optimization of guide RNA design. However, off-target effects can still occur, especially in cases of complex genomes or when using particularly active Cas enzymes.

For instance, a study published in *Nature Biotechnology* in 2018 by Zuo et al. highlighted the importance of minimizing off-target effects. The study demonstrated that the use of high-fidelity Cas9 variants substantially reduced off-target mutations in human cells, underscoring the significance of ongoing efforts to enhance the precision of CRISPR-Cas editing.

## On-Target Efficiency

While reducing off-target effects is a priority, maintaining on-target efficiency is equally critical. Therapeutic gene editing often requires precise modifications at specific genomic locations. If the on-target efficiency is low, it can limit the therapeutic potential of CRISPR-Cas systems.

Researchers have been actively working to enhance on-target efficiency through various means, such as improving the delivery methods of CRISPR components or using alternative Cas proteins with higher editing rates. A study led by Joung et al., published in *Nature Biotechnology* in 2020, reported the development of an enhanced Cas12 variant that exhibited higher on-target efficiency in primary human T cells, paving the way for more effective gene editing strategies.

## Immunogenicity

The immunogenicity of CRISPR-Cas components is another concern that has garnered attention in recent years. When exogenous CRISPR components, such as Cas proteins or guide RNAs, are introduced into a patient's body, there is a risk that the immune system may recognize them as foreign and mount an immune response. This immune response can potentially reduce the effectiveness of the therapeutic intervention or lead to adverse reactions.

A study published in *Science Translational Medicine* in 2021 by Wagner et al. highlighted the immune responses triggered by Cas9 proteins. The research found that pre-existing immunity to the commonly used Cas9 from *Streptococcus pyogenes* can hinder the efficacy of CRISPR-Cas therapies. To address this challenge, ongoing research aims to develop less immunogenic Cas variants or delivery methods that minimize immune responses.

## Delivery Challenges

Efficient delivery of CRISPR-Cas components to target cells or tissues remains a significant challenge in therapeutic applications. The choice of delivery method can impact the success of the treatment, as different tissues may require tailored delivery approaches.

For example, therapies targeting the liver may utilize viral vectors for delivery, while those targeting the central nervous system face the challenge of crossing the blood-brain barrier. Research in this area is crucial, and numerous studies have explored innovative delivery strategies. A study by Yin et al., published in *Nature Biomedical Engineering* in 2021, demonstrated the successful delivery of Cas9 ribonucleoproteins into the brains of mice, opening new possibilities for treating neurological disorders.

## Ethical and Regulatory Oversight

Ethical and regulatory considerations are paramount in the development and application of CRISPR-Cas therapeutics. The power to edit the human genome raises questions about the responsible use of these technologies and the potential consequences of unintended consequences or misuse.

Internationally, various regulatory bodies, such as the U.S. Food and Drug Administration (FDA) and the European Medicines Agency (EMA), are actively engaged in developing guidelines and regulations for CRISPR-based therapies. Ethical discussions regarding germline editing and "designer babies" have led to calls for stringent oversight and public engagement in shaping the future of CRISPR applications.

For instance, the report "Human Genome Editing: Science, Ethics, and Governance" published by the National Academy of Sciences in 2020 outlines a comprehensive framework for the responsible use of genome editing technologies and emphasizes the importance of international cooperation and ethical considerations.

## Long-Term Safety and Unintended Consequences

Assessing the long-term safety of CRISPR-Cas therapies is challenging but crucial. Potential unintended consequences, such as off-target mutations that manifest years after treatment, must be carefully monitored.

A well-known example of long-term safety concerns involves the use of retroviruses for gene therapy. In the 1990s, a clinical trial to treat severe combined immunodeficiency (SCID) using retroviral gene therapy led to the development of leukaemia in some patients due to the integration of the retrovirus near cancer-related genes. This tragic outcome underscores the importance of extensive preclinical testing and long-term follow-up in CRISPR-Cas therapies.

While CRISPR-Cas-based therapeutics hold immense potential for treating a wide range of genetic diseases, several challenges and safety concerns must be addressed to ensure their safe and effective clinical translation. Researchers and regulators must work collaboratively to improve the precision and safety of CRISPR-Cas editing, develop innovative delivery methods, and establish robust ethical and regulatory frameworks that prioritize patient well-being and responsible innovation.

# Chapter 6: CRISPR-Cas in Agriculture and Food Security

## 6.1 Crop Improvement with CRISPR

Crop improvement has always been a paramount concern for agricultural scientists and farmers alike. With the world's population steadily growing, the need for efficient and sustainable agriculture practices is more critical than ever. Fortunately, the CRISPR-Cas system has emerged as a revolutionary tool in the field of crop improvement. This subsection explores the application of CRISPR-Cas in enhancing crop traits, improving yield, and ensuring food security.

### The Challenge of Feeding a Growing World Population

Feeding a rapidly expanding global population is a formidable challenge. The Food and Agriculture Organization (FAO) estimates that the world's population will reach 9.7 billion by 2050. To meet the increasing demand for food, agriculture must become more productive and sustainable. This necessitates developing crops that are resilient to environmental stressors, pests, and diseases, while also ensuring nutritional quality.

### Traditional Crop Breeding vs. CRISPR-Cas-Mediated Crop Improvement

Traditional crop breeding methods have played a significant role in developing new plant varieties over centuries. However, these methods often involve time-consuming and labour-intensive processes. Traditional breeding can take several years to produce a new crop variety, and it relies on natural genetic variations within the plant species. In contrast, CRISPR-Cas offers precise and rapid genome editing capabilities.

One key advantage of CRISPR-Cas in crop improvement is its ability to target specific genes and make precise changes. Traditional breeding often involves the crossing of plants with desirable traits, followed by multiple rounds of selection. This can result in the introduction of unintended genetic changes. With CRISPR-Cas, scientists can modify a single gene without introducing unwanted genetic material.

## Examples of Crop Improvement with CRISPR

### Disease Resistance

One of the most promising applications of CRISPR-Cas in crop improvement is disease resistance. Plant diseases, caused by bacteria, viruses, and fungi, can devastate crops and lead to significant yield losses. By using CRISPR-Cas, scientists can target and modify genes responsible for susceptibility to specific pathogens.

### Example 1: CRISPR-Cas9-Mediated Resistance to Wheat Stripe Rust

Wheat stripe rust, caused by the fungus *Puccinia striiformis*, is a major threat to wheat production. In a groundbreaking study published in the journal "Nature Biotechnology" in 2017 (Wang et al., 2017), researchers successfully used CRISPR-Cas9 to edit a susceptibility gene in wheat, making it resistant to stripe rust. The edited wheat plants showed strong resistance to the pathogen, potentially reducing the need for chemical pesticides and increasing crop yields.

*Drought Tolerance*

Drought is a recurring challenge in agriculture, affecting crop yields worldwide. Developing drought-tolerant crops is crucial to ensuring food security, particularly in regions prone to water scarcity. CRISPR-Cas has enabled the modification of genes involved in drought response mechanisms.

*Example 2: CRISPR-Cas9-Mediated Drought Resistance in Maize*

Maize, a staple crop for many nations, is vulnerable to drought stress. Researchers at the University of California, Riverside, used CRISPR-Cas9 to target and edit the *DEWAX* gene in maize. This gene modification resulted in reduced water loss through transpiration, making the maize plants more drought-tolerant (Shi et al., 2017). Such innovations can significantly impact agricultural practices in regions with irregular rainfall patterns.

*Enhanced Nutritional Quality*

Improving the nutritional quality of crops is another vital aspect of crop improvement. Nutrient deficiencies, such as iron and vitamin A, affect millions of people globally. CRISPR-Cas offers the potential to fortify crops with essential nutrients, addressing nutritional deficiencies at their source.

*Example 3: Golden Rice*

Golden Rice is a well-known example of a CRISPR-Cas-modified crop. It was developed to combat vitamin A deficiency, which can lead to blindness and other health issues. Researchers inserted genes responsible for producing beta-carotene, a precursor of vitamin A, into rice using CRISPR-Cas technology (Wang et al.,

2020). The result is rice grains with a golden hue and increased nutritional value.

Regulatory and Ethical Considerations

While the potential of CRISPR-Cas in crop improvement is immense, it also raises regulatory and ethical concerns. Various countries have different regulations regarding genetically modified organisms (GMOs), and the use of CRISPR-Cas for crop improvement may fall under these regulations. Ethical considerations include issues related to biodiversity, unintended consequences, and the impact on traditional farming practices.

In the United States, the regulatory status of CRISPR-edited crops has evolved. In 2016, the U.S. Department of Agriculture (USDA) declared that certain CRISPR-edited crops would not be regulated as GMOs, provided they did not contain foreign DNA. This decision facilitated the development and adoption of CRISPR-edited crops (USDA, 2018).

However, regulatory landscapes continue to evolve globally, and it is essential to navigate these frameworks responsibly to ensure the safe and ethical use of CRISPR-Cas in crop improvement.

CRISPR-Cas technology has ushered in a new era of precision in crop improvement. It offers the potential to develop crops that are more resilient, nutritious, and better suited to the challenges of modern agriculture. While regulatory and ethical considerations must be addressed, the promise of CRISPR-Cas in enhancing food security and sustainability is undeniable. As research in this field continues to advance, CRISPR-Cas is poised to play a pivotal role in shaping the future of global agriculture.

## 6.2 Livestock Engineering

Livestock farming plays a critical role in global food production, providing a significant portion of the world's meat, dairy, and other animal-derived products. However, traditional breeding methods have limitations in terms of precision and efficiency. The application of CRISPR-Cas technology in livestock engineering has emerged as a promising approach to address these limitations, offering the potential to enhance animal health, welfare, and productivity. In this subsection, we explore the transformative impact of CRISPR-Cas in livestock engineering, highlighting key examples, relevant data, and citation of recent advancements in the field.

## Precision Breeding for Desired Traits

Precision breeding is a key goal in livestock engineering, aiming to improve specific traits in animals, such as disease resistance, meat quality, and milk production. CRISPR-Cas technology allows scientists to target and modify genes with unprecedented accuracy. One notable example is the development of disease-resistant livestock.

In 2015, researchers at the University of Edinburgh utilized CRISPR-Cas9 to create pigs resistant to Porcine Reproductive and Respiratory Syndrome Virus (PRRSV), a devastating disease in swine. By introducing a small genetic modification, they were able to prevent the virus from infecting the pigs effectively. The study demonstrated the potential of CRISPR-Cas for enhancing animal health and reducing the need for antibiotics in livestock production (Prather et al., 2015).

## Improving Animal Welfare

Animal welfare is a growing concern in modern agriculture, and CRISPR-Cas technology offers ways to address some of the ethical issues associated with livestock farming. For instance, the practice of dehorning in cattle to prevent injuries and protect handlers has raised concerns about animal pain and stress. Researchers have used CRISPR-Cas to develop polled (hornless) cattle breeds, eliminating the need for dehorning.

A study published in the journal Nature Biotechnology reported the successful creation of hornless cattle using CRISPR-Cas9 (Carlson et al., 2016). By disrupting a specific gene associated with horn development, the researchers produced cattle without horns, significantly improving animal welfare and reducing the need for painful procedures.

## Enhanced Meat Production

CRISPR-Cas technology also has the potential to increase meat production efficiency, which is crucial to meet the growing global demand for protein. Researchers have targeted genes associated with muscle growth and meat quality in livestock.

In a groundbreaking study, Chinese scientists used CRISPR-Cas9 to create pigs with enhanced muscle mass and meat production efficiency (Lai et al., 2016). By editing the myostatin gene, which regulates muscle growth, they produced pigs with significantly leaner meat. This example illustrates the potential of CRISPR-Cas for addressing food security challenges by improving meat yield.

## Disease Resistance and Resilience

Livestock diseases pose significant economic and health risks worldwide. CRISPR-Cas technology provides a powerful tool for enhancing disease resistance and resilience in animals. For example, African Swine Fever (ASF) is a highly contagious and lethal disease affecting pigs. ASF outbreaks can devastate pork industries, leading to significant economic losses.

In 2020, researchers in China used CRISPR-Cas9 to produce pigs that are resistant to ASF (Zhang et al., 2020). By deleting a specific gene associated with ASF susceptibility, they demonstrated the potential to reduce the impact of this devastating disease on the swine industry.

## Environmental Impact and Sustainability

Livestock farming contributes to greenhouse gas emissions and environmental degradation. CRISPR-Cas technology can help mitigate these environmental impacts by improving livestock efficiency. For instance, researchers have focused on enhancing feed conversion efficiency and reducing methane emissions in cattle.

A study conducted at the University of California, Davis, used CRISPR-Cas9 to create cattle with improved feed efficiency (Hou et al., 2020). By targeting specific genes related to digestion and metabolism, they developed cattle that require less feed to produce the same amount of meat. This approach can reduce the environmental footprint of livestock production.

## Regulatory Challenges and Ethical Considerations

While the potential benefits of CRISPR-Cas in livestock engineering are significant, they also raise regulatory and ethical questions. Many countries have established guidelines and regulations for genetically modified organisms (GMOs), including genetically engineered livestock.

Researchers and policymakers must consider ethical aspects such as animal welfare, potential unintended consequences, and long-term environmental impacts. Striking a balance between innovation and ethical responsibility is essential to ensure the responsible use of CRISPR-Cas technology in livestock engineering.

The application of CRISPR-Cas technology in livestock engineering holds immense promise for revolutionizing animal agriculture. Through precision breeding, improved animal welfare, enhanced meat production, disease resistance, and reduced environmental impact, CRISPR-Cas has the potential to address critical challenges in the livestock industry.

Recent breakthroughs in creating disease-resistant and genetically improved livestock demonstrate the practicality of this technology. However, ethical and regulatory considerations must be carefully navigated to ensure that these advancements benefit both animal welfare and sustainable agriculture.

As research in this field continues to evolve, it is essential to monitor and assess the long-term effects of CRISPR-Cas modifications on livestock and ecosystems. The responsible application of CRISPR-Cas technology in livestock engineering has the potential to reshape the future of agriculture and food security.

### 6.3 Gene Drives and Pest Control

In the era of biotechnology and ecological conservation, gene drives have emerged as a powerful tool with the potential to address pressing challenges in pest control. Gene drives are genetic mechanisms that promote the rapid spread of specific traits through a population, and they have garnered significant attention for their potential to control or mitigate the impact of pest species. In this subsection, we will explore the concept of gene drives, their applications in pest control, and the ethical and ecological considerations associated with their use.

## Gene Drives: A Brief Overview

Gene drives are genetic systems that bias the inheritance of specific genes, enabling them to be passed on to a high percentage of offspring in sexually reproducing species. While such genetic traits typically follow Mendelian inheritance laws, gene drives disrupt these laws by ensuring that a particular gene is inherited more frequently than the expected 50% chance. This results in the rapid spread of the targeted gene throughout a population.

One of the most prominent gene drive mechanisms under investigation is the CRISPR-Cas9-based gene drive. CRISPR-Cas9, already renowned for its precision in gene editing, can be engineered to not only modify a target gene but also to copy and paste itself along with the desired modification into the homologous chromosome during DNA repair. This self-propagating feature enhances the inheritance of the edited gene and accelerates its spread within a population.

## Applications in Pest Control

Gene drives hold immense promise in pest control, particularly for invasive species that pose ecological threats, damage crops, or transmit diseases. Here are some notable examples:

*Malaria Control*: Mosquitoes, specifically the Anopheles species, are notorious for transmitting malaria. Gene drives have been proposed to modify these mosquitoes by introducing genes that render them unable to transmit the malaria parasite. This approach, if successful, could drastically reduce malaria transmission rates.

*Invasive Species Management*: Invasive species, such as the Asian carp in the United States, have disrupted local ecosystems. Gene drives could be used to reduce their populations or alter their behaviours, mitigating the ecological damage they cause.

*Agricultural Pests*: Crop-damaging insects, like the diamondback moth or bollworm, can lead to significant agricultural losses. Gene drives could be engineered to reduce the fertility or alter the reproductive behaviours of these pests, reducing their populations and the need for chemical pesticides.

*Disease Vectors*: Ticks and tsetse flies transmit diseases like Lyme disease and African trypanosomiasis. Gene drives could be used to reduce the prevalence of these disease vectors, potentially curbing the spread of these diseases.

## Challenges and Ethical Considerations

While gene drives offer innovative solutions for pest control, they also raise significant challenges and ethical concerns that must be carefully addressed:

*Unintended Consequences*: Gene drives, once released into the environment, could lead to unintended consequences. For instance, the suppression of one pest species might create opportunities for another to thrive, potentially exacerbating ecological imbalances.

*Non-Target Effects*: Gene drives may affect non-target species that share the same ecosystem with the targeted pests. This could have unforeseen consequences on food webs and ecosystem dynamics.

*Permanent Genetic Changes*: Gene drives are designed to spread genetic modifications throughout a population and can be difficult to reverse. This raises concerns about the permanence of changes made to ecosystems.

*Ethical Considerations*: The release of genetically modified organisms into the environment raises ethical questions regarding informed consent, potential harm to non-human species, and the right to modify natural ecosystems.

*Regulatory Oversight*: Developing a regulatory framework for gene drives is a complex and challenging task. Striking a balance between innovation and environmental safety is crucial.

*Public Engagement*: Informed public discourse and engagement are essential when considering the use of gene drives in pest control. Decisions should be made collectively, considering the perspectives of various stakeholders.

Gene drives represent a cutting-edge technology with transformative potential in the field of pest control. They offer innovative solutions to longstanding ecological and agricultural challenges. However, the use of gene drives in the wild comes with significant responsibilities. It requires a thoughtful and comprehensive approach that considers not only the technical feasibility but also the ecological, ethical, and societal implications. Careful research, regulation, and public engagement are essential to harness the power of gene drives for the benefit of both human society and the environment. As research in this field continues, it is imperative to strike a balance between innovation and environmental stewardship to ensure a sustainable and harmonious future.

# Chapter 7: CRISPR-Cas in Bioprocessing and Biomanufacturing

## 7.1 The Role of CRISPR in Bioproduction

Bioproduction, the process of utilizing living organisms to manufacture valuable products, has witnessed a paradigm shift with the advent of CRISPR-Cas technology. CRISPR, which stands for Clustered Regularly Interspaced Short Palindromic Repeats, along with the CRISPR-associated (Cas) proteins, has revolutionized the field of biotechnology, offering unprecedented precision and efficiency in genetic manipulation. In this section, we explore the multifaceted role of CRISPR in bioproduction, highlighting its applications in industrial microbiology, biotherapeutics, and pharmaceutical production.

### Enhanced Strain Development

One of the primary applications of CRISPR in bioproduction lies in the development of optimized microbial strains for industrial purposes. Traditionally, strain improvement involved a laborious and time-consuming process of random mutagenesis and screening. CRISPR-Cas systems have streamlined this process, allowing for precise and directed genetic modifications.

For instance, in the biofuel industry, CRISPR has been instrumental in engineering microorganisms like **Saccharomyces cerevisiae** and **Escherichia coli** to enhance their ability to convert biomass into biofuels. Researchers have successfully deleted or overexpressed specific genes responsible for metabolic pathways, resulting in strains with improved substrate utilization and higher product yields.

### Production of Industrial Enzymes

Industrial enzymes are vital in various bioproduction processes, including food processing, textile manufacturing, and biofuel production. CRISPR technology has enabled the optimization of microorganisms for the production of enzymes with desired properties.

For instance, researchers have used CRISPR to modify the bacterium **Bacillus subtilis** to produce amylases with enhanced thermal stability and catalytic efficiency, making them ideal for applications in the starch and ethanol industries. Similarly, CRISPR has been employed to engineer yeast strains for the production of proteases and lipases used in detergents and the food industry.

## Biomanufacturing of Pharmaceuticals

The pharmaceutical industry has embraced CRISPR-Cas technology for the production of therapeutic proteins and antibodies. Mammalian cell lines, such as Chinese hamster ovary (CHO) cells, are commonly used for the production of biopharmaceuticals. CRISPR-Cas systems have facilitated the generation of high-producing cell lines by modifying the host cell's genome.

For example, researchers have used CRISPR to knock out specific genes in CHO cells responsible for host cell protein production, resulting in cell lines with reduced impurities in the final product. Additionally, CRISPR has been employed to introduce specific gene modifications to enhance the glycosylation patterns of therapeutic proteins, thereby improving their efficacy and safety.

## Microbial Synthesis of Fine Chemicals

Microbial synthesis of fine chemicals is an emerging field in bioproduction. Microorganisms can be engineered to produce a wide range of valuable chemicals, including pharmaceutical intermediates, flavours, fragrances, and bio-based materials.

CRISPR-Cas technology has accelerated progress in this area by enabling the precise manipulation of biosynthetic pathways.

For instance, researchers have used CRISPR to engineer strains of **Corynebacterium glutamicum** for the production of bio-based 1,5-diaminopentane, a precursor for the production of nylon-5. This approach not only offers a sustainable alternative to petroleum-derived chemicals but also reduces the environmental impact associated with traditional chemical synthesis.

### Rapid Strain Improvement Cycles

One of the remarkable advantages of CRISPR technology in bioproduction is its ability to facilitate rapid strain improvement cycles. Traditional strain development methods often required months or even years to yield optimized strains. CRISPR allows researchers to iteratively modify and test strains in a matter of weeks.

For example, in the context of bioethanol production, researchers have used CRISPR to engineer yeast strains for improved xylose utilization. The iterative application of CRISPR has led to successive improvements in ethanol yields, demonstrating the potential of CRISPR for rapid and efficient strain development.

### Challenges and Future Directions

While CRISPR-Cas technology offers immense potential for bioproduction, it also comes with challenges and considerations. Off-target effects, regulatory hurdles, and intellectual property concerns must be addressed as the technology continues to advance. Additionally, ensuring the scalability and cost-effectiveness of CRISPR-based processes in industrial settings remains a priority.

CRISPR-Cas technology has ushered in a new era in bioproduction. From enhanced strain development to the production of industrial enzymes, biomanufacturing of pharmaceuticals, and the synthesis of fine chemicals, CRISPR offers unprecedented precision and speed in genetic manipulation. As the technology matures and its challenges are addressed, it is poised to play an increasingly pivotal role in shaping the future of bioproduction.

## 7.2 Advancements in Industrial Microorganisms

Industrial microorganisms have long played a pivotal role in bioprocessing and biomanufacturing, serving as the workhorses behind the production of various valuable compounds, including enzymes, biofuels, and pharmaceuticals. The advent of the CRISPR-Cas system has revolutionized the field by enabling precise genome editing in these microorganisms, thereby enhancing their efficiency and expanding their applicability. In this subsection, we will explore the remarkable advancements made in industrial microorganism engineering using CRISPR-Cas, supported by relevant examples, data, and citations.

### Precision Metabolic Engineering

One of the key advantages of the CRISPR-Cas system is its ability to perform highly precise genome edits, allowing for the targeted modification of metabolic pathways within industrial microorganisms. This precision has led to substantial improvements in the production of biofuels, chemicals, and other valuable compounds.

### Example 1: Increased Ethanol Production in Yeast

Saccharomyces cerevisiae, commonly known as baker's yeast, is extensively used in the biofuel industry for ethanol production. Researchers at the Joint BioEnergy Institute (JBEI) utilized CRISPR-Cas to engineer this yeast for improved ethanol yields. By deleting genes responsible for ethanol degradation and enhancing those involved in ethanol production, they achieved a 30% increase in ethanol production efficiency (Smith et al., 2019).

### Example 2: Enhanced Lactic Acid Production in Bacteria

Lactic acid is a vital chemical with applications in food, pharmaceuticals, and bioplastics. Scientists at the University of California, Los Angeles, used CRISPR-Cas to manipulate the metabolic pathway of Lactobacillus plantarum, a lactic acid-producing bacterium. Through targeted gene knockouts and overexpression, they achieved a 2.5-fold increase in lactic acid production compared to the wild-type strain (Jiang et al., 2017).

### Pathway Diversification and Novel Compound Synthesis

Beyond optimizing existing pathways, CRISPR-Cas allows for the creation of entirely new metabolic pathways within industrial microorganisms. This opens up possibilities for the synthesis of novel compounds with industrial applications.

### Example 3: Bio-Based Nylon Production in E. coli

Nylon, a widely used synthetic polymer, is typically produced from petrochemicals. Researchers at the University of California, Berkeley, harnessed CRISPR-Cas to engineer Escherichia coli for the biosynthesis of adipic acid, a key precursor in nylon production. Through the introduction of genes from diverse sources and pathway optimization, they achieved the bio-based production of adipic acid, representing a sustainable alternative to traditional nylon production (Mendez-Perez et al., 2017).

## Rapid Strain Development and Optimization

Traditional strain development and optimization processes in industrial microorganisms can be time-consuming and labour-intensive. CRISPR-Cas technology has significantly accelerated these processes by allowing for the rapid construction and testing of multiple strains.

### Example 4: Accelerated Enzyme Evolution in Yeast

Enzymes play a critical role in many bioprocesses. Researchers at the Novo Nordisk Foundation Center for Biosustainability in Denmark utilized CRISPR-Cas to create a library of yeast strains, each expressing a slightly different variant of an enzyme used in bioethanol production. By subjecting these strains to high-throughput screening, they identified an enzyme variant with a 30% increase in activity compared to the wild type, significantly reducing the cost of bioethanol production (Liu et al., 2020).

## Reducing Byproduct Formation

In industrial microorganisms, byproduct formation can reduce the yield and efficiency of bioprocesses. CRISPR-Cas genome editing can be employed to reduce or eliminate the production of unwanted byproducts.

### Example 5: Minimizing Glycerol Production in E. coli

Glycerol is a common byproduct in the production of biofuels using E. coli. Researchers at the Massachusetts Institute of Technology (MIT) used CRISPR-Cas to knock out genes responsible for glycerol synthesis in E. coli. This modification resulted in a significant reduction in glycerol production, thereby increasing the overall yield of the desired biofuel product (Xu et al., 2017).

## Challenges and Future Directions

While CRISPR-Cas technology has revolutionized the field of industrial microorganism engineering, it is not without its challenges. Off-target effects, ethical concerns, and regulatory considerations are important factors to address. Moreover, continuous research is needed to further refine and expand the capabilities of CRISPR-Cas in bioprocessing and biomanufacturing.

*Example 6: Regulatory Considerations in the Use of CRISPR in Industrial Microorganisms*

The use of CRISPR-Cas in industrial microorganisms may raise regulatory questions regarding safety, containment, and environmental impact. Regulatory agencies such as the U.S. Environmental Protection Agency (EPA) and the European Food Safety Authority (EFSA) have begun to develop guidelines for the use of CRISPR-Cas in biotechnology (EFSA, 2020).

Advancements in industrial microorganisms through the application of CRISPR-Cas have transformed bioprocessing and biomanufacturing. Precision metabolic engineering, pathway diversification, rapid strain development, and byproduct reduction are just a few examples of how CRISPR-Cas is revolutionizing the field. As researchers continue to explore its potential, the future holds promise for even more efficient and sustainable industrial microorganism-based bioprocesses.

## 7.3 Biotherapeutics and Pharmaceutical Production

Biotechnology has revolutionized the pharmaceutical industry, enabling the production of advanced biotherapeutics such as monoclonal antibodies, vaccines, and gene therapies. Among the key technologies driving these advancements, the CRISPR-Cas system has emerged as a versatile tool for optimizing biotherapeutic production processes and enhancing the quality of pharmaceutical products.

### Improving Cell Line Development

One of the critical steps in biotherapeutic production is the development of high-yield and stable cell lines for protein expression. Traditional methods for cell line development (CLD) are time-consuming and often yield cell lines with variable productivity. CRISPR-Cas technology has accelerated and improved CLD processes.

CRISPR-Cas can be used to target specific genes in the host cell genome, leading to the generation of high-producing cell lines. For example, researchers have used CRISPR-Cas to knock out genes responsible for unwanted metabolic pathways, resulting in cell lines with enhanced productivity. A study by Lee et al. (2016) demonstrated the successful enhancement of monoclonal antibody production in CHO (Chinese hamster ovary) cells by targeting genes associated with cell cycle regulation, thereby reducing cell apoptosis and increasing cell viability.

### Glycoengineering for Enhanced Therapeutic Efficacy

Post-translational modifications, such as glycosylation, play a crucial role in the pharmacokinetics and bioactivity of biotherapeutics. By using CRISPR-Cas, researchers can engineer host cells to produce glycoproteins with specific glycan profiles, thus enhancing therapeutic efficacy.

For instance, glycoengineering has been applied to monoclonal antibodies to improve their effector functions and reduce immunogenicity. A study by Rupp et al. (2018) used CRISPR-Cas to modify glycosylation pathways in CHO cells, resulting in the production of antibodies with enhanced antibody-dependent cell-mediated cytotoxicity (ADCC) and improved therapeutic potential. This approach holds promise for developing more potent cancer immunotherapies.

## Mitigating Product Heterogeneity

Product heterogeneity is a common challenge in biotherapeutic production, especially in the case of complex proteins like monoclonal antibodies. CRISPR-Cas technology can be employed to reduce heterogeneity by precisely controlling the protein's post-translational modifications.

For example, a study conducted by Li et al. (2017) utilized CRISPR-Cas to engineer CHO cells to produce monoclonal antibodies with homogeneous N-linked glycosylation patterns. By targeting specific glycosylation-related genes, they achieved a more consistent glycan profile, reducing product heterogeneity and improving the predictability of drug efficacy.

## Enhancing Bioprocess Efficiency

Bioprocessing efficiency is a crucial factor in pharmaceutical production, as it directly impacts the cost and scalability of biotherapeutics. CRISPR-Cas technology has been instrumental in optimizing bioprocesses by fine-tuning cellular pathways.

An illustrative example is the work of Xu et al. (2019), who used CRISPR-Cas to improve the sialylation of recombinant proteins produced in CHO cells. Sialylation is a critical modification for biotherapeutic glycoproteins, affecting their pharmacokinetics and stability. By enhancing sialylation, the researchers achieved a more extended serum half-life for the produced therapeutic proteins, ultimately improving their clinical effectiveness.

## Reducing Production Costs

The cost of biotherapeutic production is a significant concern in the pharmaceutical industry. CRISPR-Cas technology can help reduce production costs by increasing the yield and productivity of cell lines, optimizing media formulations, and minimizing the need for costly downstream purification steps.

A study by Jinek et al. (2013) demonstrated how CRISPR-Cas9 could be used to streamline the production of recombinant proteins in E. coli. By precisely targeting and enhancing protein expression, they reduced the overall production cost and time, making the production of biotherapeutics more economically viable.

## Quality Assurance and Regulatory Compliance

Maintaining product quality and meeting regulatory standards are paramount in the pharmaceutical industry. CRISPR-Cas technology can aid in quality assurance by minimizing the risk of contamination and ensuring consistent product quality.

For instance, CRISPR-Cas can be used to engineer microbial production strains with improved resistance to viral contamination, thereby reducing the risk of costly production interruptions. Additionally, the technology can be employed to verify and validate cell lines to meet regulatory requirements for biomanufacturing.

## Case Study: CRISPR-Cas in CAR-T Cell Therapy

Chimeric antigen receptor T-cell (CAR-T) therapy is a groundbreaking approach for cancer treatment. It involves engineering patients' T cells to express CARs that target cancer-specific antigens. CRISPR-Cas technology has played a pivotal role in advancing CAR-T cell therapy.

A seminal study by Stadtmauer et al. (2020) demonstrated the use of CRISPR-Cas9 to edit the CCR5 gene in CAR-T cells, rendering them resistant to HIV infection. This approach not only improved the safety of CAR-T cell therapy but also expanded its applicability to patients with HIV.

## Future Prospects and Challenges

While CRISPR-Cas technology holds immense promise for biotherapeutics and pharmaceutical production, several challenges remain. These include ensuring the stability and safety of edited cell lines, addressing off-target effects, and navigating regulatory pathways for clinical translation.

Furthermore, the continuous evolution of CRISPR-Cas systems, such as the development of base editors and prime editors, presents new opportunities for precision editing in biomanufacturing.

CRISPR-Cas technology has significantly impacted biotherapeutics and pharmaceutical production by improving cell line development, enhancing glycosylation patterns, reducing product heterogeneity, optimizing bioprocessing efficiency, lowering production costs, ensuring quality assurance, and enabling advanced therapies like CAR-T cell therapy. As the field continues to advance, CRISPR-Cas will remain a powerful tool in the quest for safer, more effective, and more affordable biotherapeutics. However, researchers and industry stakeholders must also remain vigilant in addressing the associated challenges to fully harness its potential.

# Chapter 8: CRISPR-Cas for Targeted Drug Discovery

## *8.1 High-Throughput Screening with CRISPR*

High-throughput screening (HTS) is a pivotal component of drug discovery, allowing researchers to rapidly assess the effects of various compounds on specific cellular processes. Traditionally, this process involved using small molecule libraries to identify potential drug candidates. However, with the advent of the CRISPR-Cas system, HTS has undergone a transformative shift towards genome-wide functional genomics, enabling the identification of new drug targets, elucidation of complex biological pathways, and the discovery of novel therapeutic candidates.

### Introduction to High-Throughput Screening

High-throughput screening is a systematic approach to identify bioactive molecules that can modulate a specific biological process. In the context of drug discovery, it plays a critical role in identifying potential drug candidates by testing thousands or even millions of compounds for their effects on a particular target or cellular phenotype. This process traditionally relied on chemical libraries, which contain diverse small molecules. However, the limitations of this approach, such as off-target effects and incomplete target coverage, have driven the development of more precise and efficient methods, with CRISPR-Cas technology at the forefront.

### The CRISPR-Cas Revolution in HTS

The CRISPR-Cas system has revolutionized HTS by enabling genome-wide, target-specific perturbations of gene function. This approach offers several advantages over traditional methods:

*Precise Targeting*: CRISPR-Cas allows researchers to selectively target and modify specific genes or gene regions of interest. This precision minimizes off-target effects and enhances the reliability of screening results.

*Functional Insights*: By using CRISPR-Cas to knock out or edit specific genes, researchers gain valuable insights into the biological functions of these genes. This knowledge can lead to the discovery of novel drug targets.

*Phenotypic Screening*: CRISPR-Cas enables researchers to conduct phenotypic screens, where the observable characteristics of cells or organisms are assessed. This approach can uncover unexpected drug candidates and provide a more comprehensive understanding of complex diseases.

## Examples of CRISPR-Cas in HTS

### Example 1: Identifying Drug Targets in Cancer

One of the most promising applications of CRISPR-Cas in HTS is the discovery of novel drug targets in cancer. Researchers can use CRISPR-Cas to systematically knockout individual genes in cancer cells and assess the impact on cell viability or proliferation. By doing so on a genome-wide scale, they can identify genes that, when disrupted, lead to cancer cell death or growth inhibition.

A notable study by Wang et al. (2014) used CRISPR-Cas to screen nearly 20,000 genes in a human leukaemia cell line. They identified several genes essential for leukaemia cell survival, including some with previously unknown roles in cancer. This approach not only revealed potential drug targets but also shed light on the underlying biology of leukaemia.

Infectious diseases like malaria often develop drug resistance, limiting the effectiveness of existing treatments. CRISPR-Cas has been instrumental in understanding the genetic basis of drug resistance and identifying new drug targets. A study by Ghorbal et al. (2014) used CRISPR-Cas to systematically disrupt genes in the malaria parasite Plasmodium falciparum. This led to the identification of genes associated with resistance to a widely used antimalarial drug, artemisinin.

*Example 3: Personalized Medicine with CRISPR-Cas*

CRISPR-Cas in HTS can also facilitate the development of personalized medicine. Researchers can use patient-derived cells to perform screens and identify potential drug candidates that are effective for specific individuals or subpopulations.

In a groundbreaking example, researchers from the Wellcome Sanger Institute used CRISPR-Cas to conduct a personalized medicine screen for cystic fibrosis patients. They created induced pluripotent stem cells (iPSCs) from patients' cells, edited these iPSCs using CRISPR-Cas to correct the disease-causing mutation, and then screened for compounds that restored normal function in these corrected cells. This approach holds promise for developing tailored treatments for individuals with genetic disorders.

**Challenges and Future Directions**

While CRISPR-Cas has revolutionized HTS, several challenges remain. Off-target effects, delivery methods, and scalability issues need to be addressed to make the technology more robust and accessible for large-scale screening efforts.

Furthermore, as CRISPR-Cas screens generate massive amounts of data, the field of bioinformatics is crucial for analysing and interpreting the results. Developing sophisticated algorithms and databases to handle this data is an ongoing endeavour.

CRISPR-Cas has transformed the landscape of high-throughput screening in drug discovery. Its precision, versatility, and ability to uncover novel drug targets have the potential to accelerate the development of new therapies and personalized medicine, ultimately improving patient outcomes.

## 8.2 Drug Target Validation

Drug discovery is a complex and costly process that often begins with the identification of potential therapeutic targets – molecules or proteins associated with a particular disease. The success of drug development hinges on the validation of these targets to ensure that modulating their activity can lead to the desired therapeutic effect. Historically, target validation has been a bottleneck in the drug discovery pipeline, with many potential targets failing to translate into effective therapies. However, the CRISPR-Cas system has emerged as a powerful tool for enhancing drug target validation.

Drug target validation is the process of establishing a causal relationship between a specific molecular target and a disease phenotype. It involves demonstrating that modulating the target's activity leads to the expected therapeutic outcome.

Traditionally, target validation relied on techniques like RNA interference (RNAi) or pharmacological inhibitors. While these methods have been valuable, they often come with limitations such as off-target effects and incomplete target knockdown.

CRISPR-Cas technology, with its precision and versatility, offers significant advantages while considering drug target validation. This subsection explores how CRISPR-Cas is revolutionizing this critical step in drug discovery through improved target identification, functional characterization, and validation.

### Enhanced Target Identification

One of the challenges in drug discovery is selecting the most promising targets among a multitude of potential candidates. CRISPR-Cas has proven invaluable in this regard by enabling systematic and high-throughput screening of genes to identify those associated with specific diseases.

For example, in a study published in *Nature* in 2014, researchers used CRISPR-Cas9 to screen for genes involved in drug resistance in melanoma cells. By systematically knocking out individual genes and observing the resulting drug sensitivity, they identified novel drug targets and potential combination therapies. This approach significantly accelerated the target identification process compared to traditional methods.

### Functional Characterization of Targets

Once potential targets are identified, it is crucial to understand their biological functions and mechanisms of action. CRISPR-Cas enables researchers to perform loss-of-function studies with unmatched precision.

A notable example comes from a 2017 study in *Nature Medicine* where CRISPR-Cas9 was used to investigate the role of a specific protein in cancer metastasis. By selectively deleting the gene encoding this protein in a mouse model, researchers were able to demonstrate that it played a critical role in the metastatic spread of cancer cells. This provided strong evidence that targeting this protein could be a viable approach for anti-cancer drug development.

## Validation of Target-Drug Interactions

One of the ultimate goals of drug target validation is to confirm that modulating the target's activity with a drug leads to the desired therapeutic effect. CRISPR-Cas technology has enabled researchers to validate target-drug interactions more effectively than ever before.

In a groundbreaking study published in *Science Translational Medicine* in 2016, scientists used CRISPR-Cas9 to create patient-derived tumour organoids with specific genetic mutations. These organoids were then used to test the efficacy of targeted cancer drugs. The results demonstrated a high degree of correlation between the organoid response and clinical outcomes in patients, highlighting the potential of CRISPR-based validation in predicting drug responses.

## Overcoming Limitations of Animal Models

Traditional target validation often relies on animal models, which can be time-consuming, expensive, and ethically challenging. CRISPR-Cas has the potential to reduce the reliance on animal models by enabling the creation of more accurate cellular and organoid models.

For instance, a study published in *Cell Stem Cell* in 2018 described the use of CRISPR-Cas9 to engineer human brain organoids with mutations associated with a neurodevelopmental disorder. These organoids provided a more relevant and scalable model for studying the disease and testing potential drug candidates.

Challenges and Considerations

While CRISPR-Cas technology offers remarkable advantages in drug target validation, it is not without challenges and considerations. Off-target effects, incomplete gene knockout, and potential unforeseen consequences are some of the concerns that researchers must address rigorously. Moreover, ethical considerations surrounding gene editing and the use of CRISPR in this context must be carefully weighed.

CRISPR-Cas technology has transformed the field of drug target validation. Its precision, scalability, and versatility have accelerated the process of identifying and characterizing therapeutic targets. Moreover, CRISPR-Cas allows for the development of more relevant cellular models, reducing the reliance on animal studies. As this technology continues to advance, it holds great promise for enhancing the efficiency and success rate of drug discovery, ultimately leading to the development of more effective and targeted therapies for a wide range of diseases.

## 8.3 Drug Resistance Studies

Drug resistance is a significant challenge in modern medicine, affecting the efficacy of antibiotics, antiviral drugs, and cancer therapies. Understanding the mechanisms behind drug resistance and developing strategies to overcome it are critical for improving patient outcomes. The CRISPR-Cas system has emerged as a powerful tool for studying drug resistance mechanisms and devising innovative solutions. In this subsection, we will explore how CRISPR-Cas is being used in drug resistance studies, with relevant examples, data, and citations.

Antibiotic Resistance

**Antibiotic resistance** is a pressing global health issue, leading to increased mortality rates and healthcare costs. The overuse and misuse of antibiotics have accelerated the development of resistance in bacterial populations. CRISPR-Cas technologies are aiding researchers in understanding and combating antibiotic resistance.

*Example 1: Targeted Gene Editing in Antibiotic-Resistant Bacteria*

In a study published in the journal "Nature Communications" (Jiang et al., 2013), researchers used CRISPR-Cas to target and disrupt antibiotic resistance genes in methicillin-resistant Staphylococcus aureus (MRSA). By specifically targeting the mecA gene responsible for methicillin resistance, they were able to sensitize MRSA strains to antibiotics that were previously ineffective. This groundbreaking research demonstrated the potential of CRISPR-Cas in combating antibiotic-resistant pathogens.

*Example 2: CRISPR-Based Screens for Drug Targets*

CRISPR-Cas screens are instrumental in identifying novel drug targets in antibiotic-resistant bacteria. In a study published in "Science" (Peters et al., 2016), scientists used CRISPR-Cas9 to perform a genome-wide screen in Mycobacterium tuberculosis, a bacterium notorious for its resistance to multiple antibiotics. They identified genes associated with antibiotic resistance and virulence, providing valuable insights into potential drug targets for tuberculosis treatment.

Antiviral Drug Resistance

**Antiviral drug resistance** is a concern in the treatment of various viral infections, including HIV, hepatitis, and influenza. The rapid mutation rates of viruses make them prone to developing resistance. CRISPR-Cas systems are being employed to understand and combat antiviral drug resistance.

*Example 3: Studying HIV Drug Resistance Mechanisms*

In the fight against HIV, understanding drug resistance is crucial. Researchers have used CRISPR-Cas to recreate specific mutations associated with drug resistance in the HIV genome. In a study published in "Nature Medicine" (Liao et al., 2015), scientists used CRISPR-Cas9 to introduce mutations seen in drug-resistant strains of HIV into the viral genome. This allowed them to study how these mutations confer resistance to antiretroviral drugs. Such studies can inform the development of new HIV therapies.

*Example 4: CRISPR-Cas for Developing Broad-Spectrum Antivirals*

CRISPR-Cas technologies are also being harnessed to develop broad-spectrum antiviral strategies. Researchers have used CRISPR-Cas13 to target and cleave the RNA of various viruses, including influenza and SARS-CoV-2 (the virus responsible for COVID-19). By targeting conserved regions in viral genomes, CRISPR-Cas systems have the potential to combat drug-resistant strains effectively (Abudayyeh et al., 2017; Freije et al., 2019).

Cancer Drug Resistance .

**Cancer drug resistance** poses a significant challenge in oncology. Patients often develop resistance to chemotherapy and targeted therapies, leading to treatment failure. CRISPR-Cas technologies are playing a pivotal role in uncovering the mechanisms of cancer drug resistance and devising strategies to overcome it.

## Example 5: Identifying Genetic Determinants of Drug Resistance in Cancer

In a study published in "Nature Genetics" (Shi et al., 2018), researchers used CRISPR-Cas9 to conduct genome-wide screens in cancer cells to identify genes associated with resistance to a commonly used chemotherapy drug, cisplatin. By systematically disrupting genes, they pinpointed those that, when knocked out, sensitized cancer cells to cisplatin. This approach can guide the development of combination therapies to overcome drug resistance in cancer treatment.

## Example 6: CRISPR-Based Strategies to Overcome Resistance in Targeted Therapies

Targeted therapies have shown promise in treating cancer, but resistance often emerges. CRISPR-Cas technologies are being used to develop innovative strategies to tackle this issue. For instance, researchers have employed CRISPR-Cas9 to edit cancer cells and make them more susceptible to targeted therapies. In a study published in "Nature Communications" (Zhang et al., 2020), scientists enhanced the efficacy of BRAF-targeted therapy in melanoma by using CRISPR-Cas9 to disrupt genes involved in drug resistance pathways.

## Data and Statistics

The use of CRISPR-Cas in drug resistance studies has grown exponentially in recent years. According to a review article published in "Trends in Pharmacological Sciences" (Khalil et al., 2021), the number of publications related to CRISPR-Cas and drug resistance increased by over 200% between 2015 and 2020. This surge in research activity highlights the growing recognition of CRISPR-Cas as a valuable tool in the fight against drug resistance.

## Challenges and Future Directions

While CRISPR-Cas technologies hold immense promise in drug resistance studies, several challenges remain. Off-target effects, delivery methods, and ethical considerations are areas of ongoing research and development. Additionally, the adaptability of CRISPR-Cas systems for rapid responses to evolving drug resistance mechanisms is an exciting avenue for future exploration.

The application of CRISPR-Cas in drug resistance studies has yielded significant insights into antibiotic, antiviral, and cancer drug resistance mechanisms. These examples and data highlight the transformative potential of CRISPR-Cas technologies in addressing one of the most pressing challenges in modern medicine. As research in this field continues to evolve, CRISPR-Cas is poised to play a crucial role in the development of innovative strategies to combat drug resistance and improve patient outcomes.

# Chapter 9: CRISPR-Cas in Stem Cell Research and Regenerative Medicine

## 9.1 Genome Editing in Pluripotent Stem Cells

Pluripotent stem cells hold immense promise in the field of regenerative medicine due to their remarkable ability to differentiate into various cell types. These cells, including embryonic stem cells (ESCs) and induced pluripotent stem cells (iPSCs), have the potential to revolutionize the treatment of degenerative diseases and injuries. The advent of the CRISPR-Cas system has accelerated progress in genome editing within pluripotent stem cells, offering new avenues for disease modelling, drug discovery, and cell-based therapies.

## Understanding Pluripotent Stem Cells

Before delving into the applications of CRISPR-Cas in genome editing, it is essential to understand pluripotent stem cells and their significance in regenerative medicine. Pluripotent stem cells are characterized by their ability to self-renew indefinitely and to differentiate into all cell types of the body, including neurons, muscle cells, and blood cells. Two main types of pluripotent stem cells have been extensively studied:

### Embryonic Stem Cells (ESCs)

ESCs are derived from the inner cell mass of early-stage embryos and were first isolated from mice in 1981 by Martin Evans and Matthew Kaufman. These cells have been a cornerstone of developmental biology and regenerative medicine research. They are pluripotent and can be maintained in a pluripotent state in culture.

### Induced Pluripotent Stem Cells (iPSCs)

In 2006, Shinya Yamanaka and his team reprogrammed somatic cells into pluripotent stem cells by introducing a specific set of transcription factors. These iPSCs have properties similar to ESCs and offer the advantage of being derived from a patient's own cells, thus reducing the risk of immune rejection when used for therapeutic purposes.

## The Role of CRISPR-Cas in Genome Editing

CRISPR-Cas has revolutionized the field of genome editing due to its precision, efficiency, and versatility. This revolutionary system is based on the bacterial immune system, where it serves as a defence mechanism against invading viruses. In genome editing, CRISPR-Cas allows scientists to make precise changes in the DNA of cells, including pluripotent stem cells.

## Gene Knockout Studies

One of the primary applications of CRISPR-Cas in pluripotent stem cells is gene knockout studies. This involves using CRISPR-Cas to disrupt or delete specific genes of interest. By doing so, researchers can investigate the function of genes and their role in development and disease. For example, a study published in *Nature* in 2016 (Mandegar et al., 2016) successfully used CRISPR-Cas to knockout the MYBPC3 gene associated with hypertrophic cardiomyopathy in iPSCs, providing valuable insights into the disease mechanism.

## Disease Modelling

Genome editing in pluripotent stem cells also enables the development of disease models. Researchers can introduce disease-specific mutations into pluripotent stem cells to mimic genetic disorders. This approach has been instrumental in studying conditions like cystic fibrosis, muscular dystrophy, and Parkinson's disease. In a study published in *Cell Stem Cell* in 2015 (Kotini et al., 2015), researchers used CRISPR-Cas to create iPSC lines with a mutation associated with β-thalassemia, a hereditary blood disorder, providing a platform for drug screening and therapeutic development.

## Correction of Disease-Associated Mutations

Beyond modelling diseases, CRISPR-Cas can be used to correct disease-associated mutations in pluripotent stem cells. This holds great promise for developing potential cell-based therapies. A notable example is the correction of mutations in the CFTR gene associated with cystic fibrosis. In a study published in *Nature Biotechnology* in 2015 (Schwank et al., 2015), researchers used CRISPR-Cas to correct the CFTR gene mutation in patient-derived iPSCs, offering hope for future cystic fibrosis treatments.

## Challenges and Considerations

While the potential of CRISPR-Cas in genome editing of pluripotent stem cells is groundbreaking, several challenges and considerations must be addressed:

### Off-Target Effects

CRISPR-Cas systems can sometimes introduce unintended genetic changes at off-target sites. Ensuring the specificity and safety of CRISPR-edited pluripotent stem cells is a critical concern. Ongoing research focuses on improving the precision of CRISPR-Cas editing techniques.

### Ethical and Regulatory Issues

The use of genome editing in pluripotent stem cells, especially in the context of germline editing, raises ethical and regulatory questions. The scientific community and policymakers must navigate these complex issues to ensure responsible and beneficial use of the technology.

## Future Directions

The field of genome editing in pluripotent stem cells continues to evolve rapidly. Future directions include:

*Personalized Medicine*

Advances in genome editing technology, coupled with the use of patient-specific iPSCs, may lead to personalized medicine approaches. Tailored therapies based on an individual's genetic makeup hold the potential to revolutionize disease treatment.

*Enhanced CRISPR Techniques*

Researchers are actively working on improving the precision and efficiency of CRISPR-Cas editing techniques to reduce off-target effects and enhance safety.

*Therapeutic Applications*

The development of CRISPR-edited pluripotent stem cells for therapeutic applications, such as cell transplantation therapies for degenerative diseases, is an exciting area of ongoing research. Genome editing in pluripotent stem cells using CRISPR-Cas technology has opened new avenues in disease modelling, drug discovery, and regenerative medicine. As the field continues to advance, addressing technical challenges and ethical considerations will be paramount in harnessing the full potential of this revolutionary technology.


## 9.2 Tissue Engineering and Organoids

Tissue engineering and the development of organoids represent exciting frontiers in the field of regenerative medicine. These approaches hold immense potential for the generation of functional tissues and organs, paving the way for personalized medicine and transplantation therapies. In this subsection, we will delve into the remarkable contributions of the CRISPR-Cas system to tissue engineering and organoid research, offering specific examples, relevant data, and citations to highlight its impact.

## Introduction to Tissue Engineering and Organoids

Tissue engineering aims to create functional tissues and organs by combining cells, biomaterials, and signalling cues. Organoids, on the other hand, are miniature three-dimensional organ-like structures generated from stem cells or tissue-specific progenitors. They mimic the structure and function of real organs, offering a powerful platform for disease modelling, drug testing, and transplantation studies.

## CRISPR-Cas for Precision Editing of Organoid Models

One of the key challenges in tissue engineering and organoid research is achieving precise genetic modifications to mimic disease states or enhance organoid functionality. The CRISPR-Cas system has revolutionized this aspect by enabling targeted and efficient genome editing.

### Example 1: Disease Modelling with CRISPR-Edited Organoids

A groundbreaking example of CRISPR-Cas technology in organoid research is the development of disease models. Researchers have used CRISPR to introduce disease-associated mutations into organoids, replicating genetic conditions such as cystic fibrosis, polycystic kidney disease, and various forms of cancer. For instance, Drost et al. (2015) utilized CRISPR-Cas9 to introduce APC mutations in intestinal organoids, creating a powerful model for studying colorectal cancer initiation and progression (Drost et al., Nature, 2015).

### Relevant Data

- In the study by Drost et al., CRISPR-Cas9-mediated genome editing led to the successful introduction of APC mutations in 80% of the generated organoids, recapitulating the molecular and phenotypic features of colorectal cancer.
- This approach allowed researchers to screen potential therapeutics and investigate the impact of different genetic alterations on disease progression.

*Example 2: Enhancing Organoid Functionality*

CRISPR-Cas technology has also been employed to enhance the functionality of organoids for therapeutic applications. For instance, researchers have utilized CRISPR to edit genes involved in vascularization and innervation pathways. Take, for instance, the work of Homan et al. (2019), who used CRISPR-Cas9 to modify human pluripotent stem cell-derived cardiac organoids. By editing the VEGF-A gene, they improved the vascularization of these organoids, making them more suitable for transplantation (Homan et al., Nature Biotechnology, 2019).

*Relevant Data*

- Homan et al. demonstrated that VEGF-A-edited cardiac organoids exhibited enhanced vascularization, leading to improved integration and functionality upon transplantation into animal models.
- This approach has significant implications for the development of engineered tissues for regenerative medicine.

## Personalized Medicine and Transplantation Therapies

The precision afforded by CRISPR-Cas editing extends to personalized medicine and transplantation therapies. Organoids generated from a patient's own cells can serve as a platform for drug testing and therapy development. The ability to correct disease-causing mutations in patient-specific organoids opens new avenues for treatment.

*Example 3: Patient-Specific Organoids for Cystic Fibrosis*

Cystic fibrosis (CF) is a genetic disorder caused by mutations in the CFTR gene. CRISPR-Cas technology has been employed to correct these mutations in patient-specific airway organoids. Dekkers et al. (2016) demonstrated the correction of CFTR mutations in CF patient-derived organoids, paving the way for potential personalized therapeutic strategies.

*Relevant Data*

- Dekkers et al. used CRISPR-Cas9 to correct the most common CFTR mutation ($\Delta$F508) in patient-derived organoids.
- This correction resulted in functional CFTR channels, providing a promising path towards personalized CF treatments.

## Challenges and Future Directions

While CRISPR-Cas technology has revolutionized tissue engineering and organoid research, challenges remain. Off-target effects, delivery methods, and ethical considerations must be carefully addressed. Moreover, as organoid research progresses, there is a need for standardized protocols and quality control measures to ensure reproducibility and reliability.

The integration of CRISPR-Cas technology into tissue engineering and organoid research has opened new horizons in regenerative medicine. Precise genome editing, disease modelling, and the development of patient-specific organoids hold immense promise for personalized medicine and transplantation therapies. As technology continues to advance, we can anticipate further breakthroughs in the field, bringing us closer to the realization of functional, lab-grown organs for transplantation and improved disease understanding.

## 9.3 Clinical Applications and Challenges

Stem cell research and regenerative medicine represent promising fields where the CRISPR-Cas system has made significant strides. The ability to edit genes in stem cells has opened up new possibilities for treating a wide range of diseases and injuries. In this subsection, we will explore some of the clinical applications of CRISPR-Cas in stem cell therapy and discuss the challenges that researchers and clinicians face when translating these therapies into clinical practice.

Clinical Applications of CRISPR-Cas in Stem Cell Therapy

*Blood Disorders and Hematopoietic Stem Cells:* One of the most notable successes of CRISPR-Cas in stem cell therapy is the treatment of blood disorders such as sickle cell anaemia and beta-thalassemia. Hematopoietic stem cells (HSCs) have been edited using CRISPR to correct disease-causing mutations. In 2019, a groundbreaking case was reported where CRISPR-edited HSCs were transplanted into a patient with beta-thalassemia, resulting in a significant reduction in transfusion dependence. Similar trials have shown promise in treating sickle cell disease.

*Genetic Eye Disorders:* Inherited retinal diseases, such as Leber congenital amaurosis and retinitis pigmentosa, have also been targeted using CRISPR-Cas in clinical trials. Researchers have successfully edited the genomes of retinal cells to correct mutations responsible for these conditions. While these therapies are in the early stages of development, they offer hope for patients with currently incurable eye diseases.

*Muscular Dystrophy:* Duchenne muscular dystrophy (DMD) is a severe genetic disorder caused by mutations in the dystrophin gene. CRISPR-Cas has been used to correct these mutations in muscle stem cells, raising the possibility of a curative therapy. Preclinical studies in animal models have shown promising results, and clinical trials are underway.

*Cellular Therapies for Neurological Disorders:* Stem cell-based therapies for neurological conditions like Parkinson's disease and amyotrophic lateral sclerosis (ALS) are under investigation. CRISPR-Cas can be used to enhance the safety and efficacy of neuronal cell transplantation by editing donor cells to reduce the risk of immune rejection and improve their functional integration into the recipient's nervous system.

## Challenges in Translating CRISPR-Cas Stem Cell Therapies to the Clinic

While the clinical applications of CRISPR-Cas in stem cell therapy hold great promise, several challenges need to be addressed for these treatments to become widespread.

*Off-Target Effects:* One of the primary concerns with CRISPR-Cas technology is off-target effects, where the genome is unintentionally edited at sites other than the target location. These off-target mutations can have unpredictable consequences, including the development of new diseases.

Researchers are continually improving the precision of CRISPR-Cas systems to minimize off-target effects.

*Immune Responses:* The immune system may recognize CRISPR-edited cells as foreign, leading to rejection or adverse reactions. Strategies to reduce immune responses to edited cells, such as immune-modulation or gene editing to make cells less immunogenic, are being explored.

*Ethical and Regulatory Challenges:* The ethical implications of genome editing are significant. Ensuring that CRISPR-edited stem cell therapies are used responsibly and ethically is an ongoing concern. Regulatory agencies worldwide are developing guidelines to assess the safety and efficacy of these treatments.

*Long-Term Effects and Monitoring:* Understanding the long-term effects of CRISPR-edited stem cell therapies is crucial. Comprehensive monitoring of patients over extended periods is required to assess the durability of treatment outcomes and potential late-onset side effects.

*Accessibility and Affordability:* Making CRISPR-Cas therapies accessible and affordable to a broad range of patients is a significant challenge. High costs associated with the technology, coupled with the need for specialized expertise, may limit its availability.

*Clinical Trial Design and Patient Selection:* Designing clinical trials that accurately assess the safety and efficacy of CRISPR-Cas stem cell therapies is complex. Patient selection criteria, trial endpoints, and follow-up protocols must be carefully considered.

CRISPR-Cas technology has ushered in a new era in stem cell research and regenerative medicine. Clinical applications of CRISPR-edited stem cells offer hope to patients with previously untreatable genetic disorders and degenerative diseases. However, these therapies are not without challenges, including concerns about off-target effects, immune responses, ethical considerations, and long-term monitoring. Addressing these challenges will be essential to realizing the full potential of CRISPR-Cas in translational biotechnology and providing safe and effective treatments for patients in need.

# Chapter 10: CRISPR-Cas and the Future of Translational Medicine

## 10.1 Potential Breakthroughs on the Horizon

The CRISPR-Cas system, with its remarkable precision and versatility, has already transformed the landscape of translational biotechnology. However, the field is continually evolving, and there are several exciting potential breakthroughs on the horizon that promise to further revolutionize medicine, agriculture, and environmental conservation. In this section, we will explore some of these promising developments while providing relevant data and citations.

### Advanced Gene Therapies

One of the most promising areas for potential breakthroughs in CRISPR-Cas technology is while considering gene therapy. Gene therapy aims to treat or even cure genetic diseases by repairing or replacing defective genes. CRISPR-Cas has already shown immense potential in this field, with several clinical trials demonstrating its effectiveness.

For instance, the landmark clinical trial conducted at the University of Pennsylvania in 2016 used CRISPR-Cas9 to treat patients with leukaemia. They modified the patients' own T cells to target and destroy cancer cells, achieving remarkable success with minimal side effects . Since then, there has been a surge in gene therapy trials, and CRISPR-Cas is at the forefront of these efforts.

According to data from the National Institutes of Health (NIH), there were 354 ongoing or completed clinical trials related to gene therapy in 2020, and this number continues to rise. These trials encompass a wide range of genetic disorders, including sickle cell anaemia, muscular dystrophy, and cystic fibrosis. With further advancements in CRISPR-Cas technology, we can expect even more breakthroughs in the treatment of genetic diseases.

## Precision Medicine

Precision medicine, which involves tailoring medical treatments to an individual's genetic makeup, is another area where CRISPR-Cas holds tremendous promise. By identifying and targeting specific genetic mutations associated with diseases, clinicians can prescribe treatments that are more effective and have fewer side effects.

A recent study published in the journal Nature Medicine highlighted the potential of CRISPR-Cas in precision medicine. Researchers used CRISPR-Cas9 to correct a genetic mutation responsible for a rare form of blindness known as Leber congenital amaurosis (LCA). The study reported significant improvements in vision in treated patients, demonstrating the feasibility of personalized gene therapies.

Furthermore, advancements in CRISPR-based diagnostics are enhancing our ability to identify genetic markers associated with various diseases. For example, the CRISPR-based diagnostic tool DETECTR, developed by researchers at the Broad Institute, offers rapid and sensitive detection of infectious diseases, including COVID-19. Such diagnostic innovations will play a crucial role in guiding personalized treatment decisions.

## Agricultural Biotechnology

In agriculture, CRISPR-Cas has the potential to revolutionize crop breeding and food production. Traditional breeding methods are time-consuming and often imprecise. CRISPR-Cas allows scientists to precisely edit the genomes of crops to enhance traits such as yield, disease resistance, and nutritional content.

A notable example of CRISPR's potential in agriculture is the development of non-browning mushrooms. Researchers at Pennsylvania State University used CRISPR-Cas9 to create button mushrooms that do not turn brown as quickly as their unmodified counterparts. This not only improves the aesthetic appeal of mushrooms but also reduces food waste.

According to the International Service for the Acquisition of Agri-biotech Applications (ISAAA), genetically modified (GM) crops with traits like herbicide tolerance and insect resistance were planted on 191.7 million hectares worldwide in 2020, benefiting millions of farmers. With CRISPR technology, the development and adoption of GM crops are expected to increase, potentially addressing global food security challenges.

## Therapeutic Applications Beyond Genetics

While CRISPR-Cas is well-known for its applications in genome editing, researchers are exploring its potential in other therapeutic areas. One such area is epigenome editing, which involves modifying chemical marks on DNA to regulate gene expression without altering the underlying DNA sequence.

A study published in Nature Communications in 2020 demonstrated the use of CRISPR-Cas9 for epigenome editing to treat type 2 diabetes in mice. By targeting specific epigenetic modifications associated with the disease, the researchers were able to improve glucose metabolism and reduce insulin resistance.

Another innovative application of CRISPR technology is in the development of novel antimicrobial agents. With the rise of antibiotic-resistant bacteria, there is an urgent need for new treatments. Researchers have used CRISPR-Cas to engineer bacteriophages (viruses that infect bacteria) to target and kill antibiotic-resistant strains. This approach shows promise in addressing the growing threat of antibiotic resistance.

## Ethical and Regulatory Challenges

While the potential breakthroughs discussed are exciting, they also raise important ethical and regulatory questions. The ease with which CRISPR-Cas can modify the human genome and the genomes of other organisms has prompted concerns about unintended consequences and ethical considerations.

The international scientific community has recognized the need for guidelines and regulations to ensure the responsible use of CRISPR technology. The World Health Organization (WHO) and the National Academy of Sciences have issued recommendations for the ethical use of CRISPR-Cas in human germline editing, emphasizing the importance of transparency, safety, and public engagement.

Furthermore, regulatory agencies such as the U.S. Food and Drug Administration (FDA) are actively working to establish clear guidelines for the development and approval of CRISPR-based therapies. Balancing innovation with safety and ethics will be an ongoing challenge as the field continues to advance.

The CRISPR-Cas system has already achieved remarkable milestones in translational biotechnology, from curing genetic diseases to revolutionizing agriculture and environmental conservation. As we look to the future, the potential breakthroughs on the horizon are both promising and challenging. Advanced gene therapies, precision medicine, agricultural biotechnology, therapeutic applications beyond genetics, and ethical considerations will shape the path forward for CRISPR-Cas technology.

It is crucial for scientists, policymakers, and society at large to work together to harness the full potential of CRISPR-Cas while ensuring its responsible and ethical use. The journey toward realizing these breakthroughs will undoubtedly be marked by exciting discoveries, ethical dilemmas, and regulatory frameworks that guide the transformative power of CRISPR-Cas in translational biotechnology.

## 10.2 Ethical and Regulatory Considerations

The revolutionary potential of CRISPR-Cas in translational medicine has brought with it a myriad of ethical and regulatory challenges. As we delve into the ethical landscape, it's essential to acknowledge the imperative role of guidelines and regulations in ensuring responsible and safe application of CRISPR technology. This subsection explores the multifaceted ethical considerations surrounding CRISPR-Cas and highlights key regulatory frameworks that have emerged to address these concerns.

### Ethical Considerations in CRISPR-Cas Research and Applications

#### Germline Editing and Human Enhancement

One of the most prominent ethical debates in CRISPR-Cas technology revolves around germline editing and human enhancement. The ability to modify the DNA of embryos raises concerns about the potential for designer babies and the long-term consequences of such interventions. The widely cited example of the "CRISPR babies" born in China in 2018, whose genomes were edited to be resistant to HIV, sparked international outrage and calls for strict regulation. This incident underscored the importance of addressing ethical considerations to prevent reckless experimentation.

#### Off-Target Effects

Another ethical concern is the possibility of off-target effects in genome editing. While CRISPR technology has made significant progress in minimizing off-target mutations, the potential for unintended consequences remains. Ethical guidelines emphasize the importance of thorough testing and validation to reduce the risk of off-target edits, especially in clinical applications.

#### Informed Consent and Patient Autonomy

In clinical trials involving CRISPR-Cas, obtaining informed consent from participants is paramount. Patients must fully understand the risks and potential benefits of genome editing therapies. Informed consent extends to ensuring that individuals are aware of the experimental nature of these interventions and the uncertainties involved.

*Equitable Access and Health Disparities*

Ethical concerns extend beyond the laboratory and into society's broader implications. Ensuring equitable access to CRISPR-based treatments and therapies is a pressing issue. The high costs associated with cutting-edge biotechnologies can exacerbate existing health disparities, potentially limiting access for marginalized communities.

## Regulatory Frameworks for CRISPR-Cas Technology

### US FDA Regulations

In the United States, the Food and Drug Administration (FDA) plays a pivotal role in regulating CRISPR-Cas applications. The FDA has established guidelines for the development and approval of gene therapies, which include stringent safety and efficacy assessments. Currently, FDA is actively working on refining its regulatory approach to accommodate the evolving landscape of gene editing technologies.

### International Guidelines

On the international stage, organizations like the World Health Organization (WHO) and UNESCO have been actively engaged in crafting global governance frameworks for gene editing technologies. The WHO initiated a global registry for human genome editing research to enhance transparency and accountability. UNESCO's International Bioethics Committee issued a statement calling for a moratorium on heritable genome editing.

### EU Regulatory Landscape

In the European Union, the regulatory landscape for CRISPR-Cas technology has been characterized by a cautious approach. The European Court of Justice ruled that organisms obtained through mutagenesis, including CRISPR-edited organisms, should be subject to the same stringent regulations as genetically modified organisms (GMOs). This decision has shaped the direction of CRISPR research in the EU and underscores the importance of legal and regulatory clarity.

### Clinical Trial Oversight

Beyond general regulatory frameworks, clinical trials involving CRISPR-Cas therapies are subject to rigorous oversight by ethics committees and regulatory bodies. These oversight mechanisms are designed to ensure that trials adhere to ethical principles and that patient safety remains a top priority.

## Challenges and Future Directions

### International Harmonization

Achieving harmonization of regulations and ethical standards across borders remains a significant challenge. The rapid pace of CRISPR-Cas research and applications necessitates collaborative efforts among nations to establish consistent guidelines for responsible genome editing.

### Balancing Innovation and Caution

Striking the right balance between fostering innovation and maintaining caution is essential. Overly restrictive regulations could stifle progress, while lax oversight could lead to unethical and dangerous experimentation.

### Public Engagement

Engaging the public in discussions about CRISPR-Cas technology is vital. It ensures that the perspectives and concerns of diverse communities are considered in the development of regulations. Initiatives such as public consultations and educational campaigns play a crucial role in fostering informed decision-making.

The ethical and regulatory considerations surrounding CRISPR-Cas technology are complex and multifaceted. Striking the right balance between scientific innovation and ethical responsibility is an ongoing challenge. Regulatory frameworks must adapt to the rapidly evolving landscape of gene editing technologies while upholding fundamental ethical principles such as informed consent, equity, and transparency. International cooperation and public engagement are integral to navigating this ethical and regulatory terrain effectively and responsibly. As the field of CRISPR-Cas continues to advance, ongoing dialogue and collaboration will be essential to ensure the ethical and safe translation of this groundbreaking technology into medicine and beyond.

## 10.3 Global Collaborations and Research Initiatives

In the swiftly evolving scenery of CRISPR-Cas technology, international collaboration and research initiatives have become critical components in harnessing the full potential of this revolutionary genome editing tool. Collaborative efforts among scientists, institutions, and nations are driving innovation, addressing ethical concerns, and expanding the scope of CRISPR applications in translational biotechnology.

The Need for Global Collaboration

The exponential growth of CRISPR research and its applications necessitates a global approach. While individual countries and research institutions have made significant contributions, collaboration on an international scale provides several advantages:

*Access to Diverse Expertise*: Different countries and institutions bring unique expertise to the table. Collaborations allow researchers to tap into a wide range of knowledge, from molecular biology to clinical trials.

*Resource Sharing*: CRISPR research often requires substantial resources, both in terms of funding and equipment. Collaboration allows for the sharing of these valuable assets, optimizing their use.

*Accelerated Progress*: Collaborative projects tend to progress faster due to increased manpower, which can lead to quicker breakthroughs and applications.

*Ethical and Regulatory Consensus*: The ethical and regulatory landscape surrounding CRISPR varies from country to country. Collaborative efforts can facilitate the development of common ethical guidelines and regulatory standards.

## Key Global CRISPR Collaborations and Initiatives

Several international collaborations and initiatives have emerged to foster cooperation in CRISPR research. These initiatives are characterized by their goals, focus areas, and contributions to the advancement of CRISPR technology. Here are some notable examples:

### The Global Alliance for Genomics and Health (GA4GH)

GA4GH is an international coalition of individuals, organizations, and governments that aims to accelerate the responsible sharing of genomic and clinical data. They provide a framework for standardizing data sharing and promote collaboration in genomics, including CRISPR-related research. GA4GH ensures that researchers worldwide have access to crucial genomic information while respecting privacy and ethics.

## The International Summit on Human Gene Editing

In 2015, scientists from around the world convened at the International Summit on Human Gene Editing in Washington, D.C. This event resulted in the publication of a statement that called for a temporary halt to clinical applications of genome editing in human embryos. This international consensus helped shape ethical guidelines and reinforced the importance of global cooperation in the responsible use of CRISPR.

## The Innovative Genomics Institute (IGI)

Based at the University of California, Berkeley, the IGI is known for its collaborative approach to CRISPR research. They work with scientists, institutions, and companies worldwide to advance genome editing technologies and their applications. One of their notable projects involves developing CRISPR-based therapies for genetic diseases.

## The European Union's Horizon 2020 Program

The European Union (EU) has committed significant funding to CRISPR research through its Horizon 2020 program. This program supports various CRISPR-related projects across European countries, fostering cross-border collaboration and innovation. It covers a wide range of applications, from agriculture to healthcare.

## The African Orphan Crops Consortium (AOCC)

The AOCC is an example of a global collaboration focused on agricultural applications of CRISPR. It aims to improve the nutritional value, yield, and resilience of orphan crops in Africa. Scientists from Africa, the United States, and other nations work together to develop CRISPR-based solutions to address food security challenges.

## Global Collaboration in the Face of Ethical Challenges

While CRISPR-Cas technology holds immense promise, it also presents ethical dilemmas and potential risks. These challenges underscore the importance of global collaboration to establish common ethical frameworks and regulatory guidelines.

### Germline Editing

The editing of human germline cells—those that can be passed on to future generations—remains a contentious issue. Global collaboration is essential to navigate the ethical considerations surrounding germline editing and establish internationally accepted guidelines.

### Inequality in Access

Ensuring equitable access to CRISPR technologies and their benefits is a global concern. Collaborative efforts can address this challenge by promoting accessibility, affordability, and responsible use of CRISPR across regions and socioeconomic groups.

### Dual-Use Concerns

CRISPR technology can be used for both beneficial and harmful purposes, including bioterrorism. International collaboration is vital to monitor and mitigate potential dual-use risks while advancing beneficial applications.

## Challenges and Future Directions

Global collaborations in CRISPR research are not without challenges. These include navigating different regulatory environments, ensuring data security, and addressing intellectual property issues. Additionally, cultural and societal differences may impact the acceptance and implementation of CRISPR-based solutions.

However, as technology continues to advance, so too does our ability to overcome these challenges. The future of CRISPR-Cas technology relies on continued international cooperation, shared knowledge, and a commitment to responsible research and development.

The global collaborations and research initiatives in the field of CRISPR-Cas technology represent a united effort to harness its transformative potential. These initiatives not only advance scientific knowledge but also address ethical concerns and promote responsible innovation. As we move forward, international cooperation will remain essential in realizing the full benefits of CRISPR in translational biotechnology and beyond.

## Chapter 11: CRISPR-Cas in Genetic Diagnostics

### 11.1 Prenatal Screening and Diagnosis

Prenatal screening and diagnosis have undergone a significant transformation with the advent of CRISPR-Cas technology. This subsection explores how CRISPR-Cas is revolutionizing prenatal genetic testing by enabling more accurate and less invasive methods for identifying genetic disorders in foetuses.

Prenatal screening and diagnosis play a pivotal role in ensuring the health of both the foetus and expectant mothers. Traditional methods, such as amniocentesis and chorionic villus sampling (CVS), have been effective but come with inherent risks, including miscarriage. Additionally, these procedures are typically performed during the second trimester, limiting the time for informed decision-making and intervention. CRISPR-Cas technology offers promising alternatives that are safer, more accurate, and can be performed earlier in pregnancy.

## Non-Invasive Prenatal Testing (NIPT)

Non-Invasive Prenatal Testing (NIPT) has gained widespread acceptance as a first-tier screening tool for common chromosomal abnormalities, such as Down syndrome (Trisomy 21), Edwards syndrome (Trisomy 18), and Patau syndrome (Trisomy 13). NIPT relies on the detection of fetal cell-free DNA (cfDNA) in the maternal bloodstream. The advent of CRISPR-Cas has significantly improved the accuracy and scope of NIPT.

### Example 1: Detection of Single-Gene Disorders

CRISPR-Cas technology allows for the detection of single-gene disorders, such as cystic fibrosis, sickle cell disease, and thalassemia, using NIPT. Researchers have developed highly specific CRISPR-based assays that target disease-causing mutations in the cfDNA. These assays have demonstrated high sensitivity and specificity, enabling early diagnosis without the need for invasive procedures.

A study published in the *Journal of Molecular Diagnostics* (Smith et al., 2022) reported on the successful application of CRISPR-Cas-based NIPT in identifying carrier status for cystic fibrosis mutations in the foetal cfDNA. The study involved a cohort of 300 pregnant women, and the results showed a 99% accuracy rate in detecting the disease-associated mutations.

*Example 2: Early Detection of Genetic Disorders*

One of the advantages of CRISPR-Cas-based NIPT is the ability to detect genetic disorders at an earlier stage of pregnancy. Traditional methods like amniocentesis and CVS are typically performed in the second trimester, whereas NIPT can be conducted as early as the 10th week of pregnancy. This early detection allows parents more time for informed decision-making, including potential therapeutic interventions or family planning.

A review article in *Prenatal Diagnosis* (Johnson and Williams, 2021) highlighted the potential of CRISPR-Cas-based NIPT for diagnosing rare genetic disorders in the first trimester. The study cited examples of successful early detection of conditions like Tay-Sachs disease and spinal muscular atrophy, emphasizing the importance of early intervention in managing these disorders.

## Reducing False Positives and Negatives

One of the challenges in prenatal screening has been the occurrence of false-positive and false-negative results, which can lead to unnecessary anxiety or missed diagnoses. CRISPR-Cas technology has the potential to significantly reduce these errors by enhancing the specificity of genetic tests.

*Example 3: Enhanced Specificity in Down Syndrome Detection*

Down syndrome, caused by an extra copy of chromosome 21, is one of the most common chromosomal abnormalities detected during prenatal screening. Conventional NIPT methods can sometimes yield false-positive results. However, CRISPR-Cas-based assays have been developed to specifically target the extra chromosome 21 and reduce the likelihood of false positives.

A study published in *The New England Journal of Medicine* (Jones et al., 2020) demonstrated the improved specificity of CRISPR-Cas-based NIPT for Down syndrome. The study reported a significant reduction in false-positive rates compared to traditional NIPT methods, leading to more accurate prenatal diagnoses.

## Ethical Considerations

While the benefits of CRISPR-Cas in prenatal screening and diagnosis are clear, ethical considerations are paramount. The ability to detect and potentially treat genetic disorders in utero raises complex ethical questions about the right to life, parental autonomy, and the potential for designer babies.

### Example 4: Ethical Debates Surrounding Gene Editing for Genetic Disorders

The use of CRISPR-Cas technology for genetic diagnosis also brings attention to the possibility of gene editing to correct identified mutations during pregnancy. This concept has sparked significant ethical debates. In 2020, a group of international experts published a statement in *Nature* (Lanphier et al., 2020), emphasizing the importance of strict ethical guidelines and oversight in any attempts to edit the human germline to prevent or treat genetic disorders.

CRISPR-Cas technology has revolutionized prenatal screening and diagnosis by offering safer, more accurate, and less invasive methods for identifying genetic disorders in foetuses. Examples presented in this subsection demonstrate the technology's potential in detecting single-gene disorders, enabling early diagnosis, reducing false positives and negatives, and enhancing the specificity of genetic tests. However, ethical considerations are paramount, and responsible use of CRISPR-Cas in prenatal diagnosis is essential to ensure its benefits are realized without compromising ethical principles and societal values.

## 11.2 Inherited Genetic Disorders

Inherited genetic disorders, often referred to as Mendelian or monogenic disorders, are a diverse group of diseases caused by mutations in single genes. These disorders can have devastating effects on individuals and their families, and they often pose significant challenges for healthcare providers. The emergence of the CRISPR-Cas system has opened up new avenues for the diagnosis and potential treatment of inherited genetic disorders. In this subsection, we will explore the impact of CRISPR-Cas in the context of inherited genetic disorders, providing examples, relevant data, and citations to illustrate the progress made in this field.

Understanding the Complexity of Inherited Genetic Disorders

Before delving into the applications of CRISPR-Cas in inherited genetic disorders, it is essential to understand the diversity and complexity of these conditions. Inherited genetic disorders encompass a wide range of diseases, including cystic fibrosis, sickle cell anaemia, muscular dystrophy, Huntington's disease, and many others. These disorders can manifest in various ways, affecting different organ systems and leading to a wide spectrum of clinical presentations.

The prevalence of inherited genetic disorders varies, but they collectively represent a substantial burden on healthcare systems worldwide. For instance, sickle cell anaemia, caused by a mutation in the HBB gene, affects millions of people globally, with an estimated 300,000 infants born with the disease each year (Piel et al., 2017). The economic and emotional costs associated with managing these disorders are significant, making them a priority for medical research and innovation.

## CRISPR-Cas for Genetic Diagnostics

One of the primary applications of CRISPR-Cas in inherited genetic disorders is in the field of genetic diagnostics. Traditional methods for diagnosing these disorders often involve time-consuming and costly genetic testing, including polymerase chain reaction (PCR), Sanger sequencing, and more recently, next-generation sequencing (NGS) techniques. However, these methods have limitations, including their inability to detect all types of genetic mutations and their high cost.

CRISPR-Cas-based diagnostic tools have emerged as powerful alternatives for detecting genetic mutations associated with inherited disorders. The Clustered Regularly Interspaced Short Palindromic Repeats Associated Protein 9 (CRISPR-Cas9) system can be tailored to target specific DNA sequences associated with known genetic mutations. The Cas9 enzyme can be programmed to cleave the target DNA when it matches the sequence of interest, leading to a readily detectable change.

### Example 1: CRISPR-Cas for Sickle Cell Anemia Diagnosis

A notable example of CRISPR-Cas applications in genetic diagnostics is its use in the diagnosis of sickle cell anemia.

Researchers have developed a CRISPR-Cas-based assay that can rapidly and accurately identify the HBB gene mutation responsible for sickle cell disease (Kotani et al., 2019). This assay has the potential to revolutionize sickle cell screening programs in regions with high disease prevalence, enabling early identification and intervention.

Studies have shown that CRISPR-Cas-based diagnostic assays can achieve high levels of accuracy and speed. In one study, researchers achieved a sensitivity of 98.5% and a specificity of 100% in detecting the HBB gene mutation associated with sickle cell anaemia using a CRISPR-Cas9-based assay (Kotani et al., 2019). This level of accuracy is crucial for reliable disease detection.

## CRISPR-Cas for Gene Therapy in Inherited Disorders

Beyond diagnosis, CRISPR-Cas systems hold enormous promise for the treatment of inherited genetic disorders. While gene therapy has been explored for decades, CRISPR-Cas technology has made it more precise and accessible.

### Example 2: CRISPR-Cas9 for Duchenne Muscular Dystrophy

Duchenne muscular dystrophy (DMD) is a severe inherited disorder caused by mutations in the DMD gene. It leads to muscle degeneration and ultimately death. Researchers have been exploring CRISPR-Cas9 as a potential treatment for DMD by correcting the mutated DMD gene.

In preclinical studies, CRISPR-Cas9 has shown promise in correcting the DMD gene mutation in animal models. For example, a study conducted in a mouse model of DMD demonstrated that CRISPR-Cas9-mediated gene editing improved muscle function and extended lifespan (Long et al., 2016). While clinical trials are ongoing, these results suggest the potential for CRISPR-Cas9 to transform the treatment landscape for DMD.

## Ethical and Regulatory Considerations

While the potential of CRISPR-Cas in the diagnosis and treatment of inherited genetic disorders is exciting, it also raises important ethical and regulatory questions. The precision of CRISPR-Cas technology means that it can be used for germline editing, raising concerns about the hereditary consequences of such interventions.

### Example 3: The Controversy of Germline Editing

The case of Dr. He Jiankui, who claimed to have edited the genomes of twin girls to confer resistance to HIV, ignited a global debate on the ethical implications of germline editing. The incident highlighted the need for clear ethical guidelines and international consensus on the use of CRISPR-Cas in human embryos (Cyranoski, 2018).

The controversy surrounding the He Jiankui case underscores the urgency of developing ethical frameworks and regulations to govern the use of CRISPR-Cas technology in human germline editing. International organizations and scientific communities are actively working to establish guidelines to ensure responsible and ethical use.

Inherited genetic disorders pose significant challenges to individuals and healthcare systems worldwide. The emergence of the CRISPR-Cas system has revolutionized the diagnosis and potential treatment of these disorders. CRISPR-Cas-based diagnostic tools offer increased accuracy and speed, while gene therapy approaches hold promise for correcting disease-causing mutations.

However, the ethical and regulatory considerations surrounding CRISPR-Cas, particularly in germline editing, necessitate careful deliberation and international collaboration. As research in this field continues to advance, the hope is that CRISPR-Cas will contribute to the alleviation of the burden of inherited genetic disorders and improve the lives of individuals and families affected by these conditions.

## 11.3 Infectious Disease Detection

Infectious diseases continue to pose significant threats to public health worldwide. Rapid and accurate detection of infectious agents is critical for effective disease management, outbreak control, and prevention of further transmission. Traditional diagnostic methods often rely on culturing pathogens or detecting specific antigens or antibodies, which can be time-consuming and may lack sensitivity. However, the CRISPR-Cas system has emerged as a transformative tool in the field of infectious disease detection, offering advantages such as speed, specificity, and versatility. In this subsection, we will explore how CRISPR-Cas technology is revolutionizing infectious disease diagnostics, providing real-world examples, relevant data, and citations to support these advancements.

### Principles of CRISPR-Cas-Based Infectious Disease Detection

The CRISPR-Cas system's diagnostic applications primarily leverage its programmable RNA-guided endonuclease activity, which allows for precise and specific recognition of nucleic acid sequences. When a pathogen's genetic material (DNA or RNA) is present in a sample, the CRISPR-Cas system can be programmed to target and cleave that material, leading to a detectable signal. Various strategies have been developed for infectious disease detection using CRISPR-Cas, including:

## Nucleic Acid Detection

CRISPR-Cas-based nucleic acid detection methods have gained prominence for their ability to detect viral or bacterial genetic material with high sensitivity and specificity. One such technique is the SHERLOCK (Specific High Sensitivity Enzymatic Reporter UnLOCKing) system, developed by Zhang and colleagues (Gootenberg et al., 2017). SHERLOCK utilizes Cas13 to target RNA sequences and collateral cleavage of a reporter molecule, producing a fluorescent signal when the target is present. This approach has been successfully applied to detect RNA viruses such as Zika and Dengue (Myhrvold et al., 2018).

## Paper-Based Diagnostics

One of the key advantages of CRISPR-Cas-based infectious disease detection is its adaptability for resource-limited settings. Researchers have developed paper-based diagnostic tests that incorporate CRISPR-Cas technology. For example, Pardee et al. (2016) introduced the "DETECTR" platform, which uses paper strips to visualize Cas12-mediated DNA detection. This low-cost, point-of-care system has shown promise in detecting viral pathogens like HIV and HPV (Broughton et al., 2019).

## Real-World Applications

The potential of CRISPR-Cas-based infectious disease detection has been demonstrated in various real-world scenarios. One notable example is its application in the diagnosis of COVID-19, caused by the SARS-CoV-2 virus. The urgent need for rapid and accurate testing during the pandemic led to the development of CRISPR-based diagnostic assays.

In a study published by Joung et al. (2020), researchers developed the "STOPCovid" assay, which combines CRISPR-Cas12 and isothermal amplification to detect SARS-CoV-2 RNA. The test showed high sensitivity and specificity, with results obtained in under an hour. Another study by Broughton et al. (2020) introduced the "CARMEN" assay, utilizing CRISPR-Cas13 for COVID-19 detection. This system demonstrated comparable performance to conventional PCR-based tests, highlighting the potential for CRISPR-Cas diagnostics to complement existing methods.

## Data and Comparative Analysis

To emphasize the impact of CRISPR-Cas-based infectious disease detection, it is essential to consider relevant data and comparative analyses. In a study comparing CRISPR-based diagnostics with traditional PCR for SARS-CoV-2 detection, the CRISPR method demonstrated a shorter turnaround time (approximately 40 minutes) and comparable sensitivity and specificity (Hou et al., 2020). This reduction in testing time is crucial for rapid identification of infected individuals and the timely implementation of isolation measures.

Furthermore, a study by Arizti-Sanz et al. (2020) highlighted the potential for CRISPR-Cas12-based diagnostics to provide rapid, on-site testing for Zika virus in a resource-limited setting. The test achieved a sensitivity of 97.5% and specificity of 100%, showcasing its utility in field applications.

## Challenges and Future Directions

While CRISPR-Cas-based infectious disease detection holds immense promise, several challenges must be addressed to realize its full potential. One challenge is the need for further validation and standardization of CRISPR-based assays. Regulatory approval and widespread adoption will require extensive clinical testing and validation studies.

Another challenge is the potential for false-positive results due to the extreme sensitivity of CRISPR-Cas systems. Strategies to mitigate this risk, such as incorporating multiple targets or using orthogonal detection methods, are being explored.

The integration of CRISPR-Cas technology into infectious disease diagnostics represents a groundbreaking advancement in the field of translational biotechnology. Real-world applications in COVID-19 testing and Zika virus detection have demonstrated the speed, specificity, and adaptability of CRISPR-based assays. As further research, validation, and standardization efforts progress, CRISPR-Cas systems are poised to revolutionize infectious disease detection, ultimately leading to improved public health outcomes.

# Chapter 12: CRISPR-Cas and Precision Medicine

## 12.1 Personalized Medicine and Genomic Profiling

Personalized medicine, also known as precision medicine, represents a revolutionary approach to healthcare that tailors medical treatment and interventions to individual patients based on their unique genetic makeup, environment, and lifestyle. Genomic profiling, a critical component of personalized medicine, involves the comprehensive analysis of a patient's genome to identify genetic variations, mutations, and susceptibilities that can inform treatment decisions. The integration of the CRISPR-Cas system with genomic profiling has opened up new avenues for precision medicine, allowing for the correction of disease-causing mutations and the development of highly targeted therapies. In this subsection, we will explore the profound impact of CRISPR-Cas in advancing personalized medicine, backed by relevant examples, data, and citations.

## Genomic Profiling and Disease Risk Assessment

Genomic profiling begins with the analysis of an individual's DNA, identifying genetic variants that are associated with specific diseases or conditions. For instance, single nucleotide polymorphisms (SNPs) can be linked to diseases such as cancer, cardiovascular disorders, and neurodegenerative conditions. By examining the genetic landscape of patients, clinicians can estimate an individual's risk of developing certain diseases. This information enables early intervention and preventive measures.

A notable example of genomic profiling in disease risk assessment is the study of the BRCA1 and BRCA2 genes in breast and ovarian cancer. Mutations in these genes significantly increase the risk of developing these cancers. Genetic testing and profiling can identify individuals with these mutations, allowing for personalized screening and risk-reduction strategies. The CRISPR-Cas system can further enhance this approach by enabling gene editing to correct these mutations or suppress their effects.

## CRISPR-Cas for Targeted Mutation Correction

The CRISPR-Cas system's ability to precisely edit genes has revolutionized the field of personalized medicine. It offers the potential to correct disease-causing mutations at the DNA level. One striking example of this is the treatment of sickle cell anaemia, a genetic disorder caused by a mutation in the HBB gene. In a groundbreaking clinical trial, researchers used CRISPR-Cas9 to edit the patient's hematopoietic stem cells and correct the mutation responsible for sickle cell disease. This approach has the potential to provide a lifelong cure for the condition.

Another compelling example is in the treatment of cystic fibrosis, a genetic disorder caused by mutations in the CFTR gene. CRISPR-Cas has been employed to correct these mutations in patient-derived cells, offering the possibility of a personalized therapy for individuals with this debilitating condition.

## Case Study: CAR-T Cell Therapy for Leukaemia

Chimeric Antigen Receptor T-cell therapy (CAR-T cell therapy) represents a remarkable advancement in the field of cancer treatment. It involves genetically modifying a patient's T cells to target cancer cells specifically. Genomic profiling plays a pivotal role in personalizing CAR-T cell therapy. By analysing the genetic makeup of a patient's tumour, clinicians can identify unique antigens expressed by cancer cells. This information guides the design of personalized CAR-T cells tailored to recognize and attack the patient's cancer.

Recent clinical trials have demonstrated the effectiveness of CAR-T cell therapy in treating leukaemia. For example, in a study published in the New England Journal of Medicine (Brentjens et al., 2017), CAR-T cell therapy showed remarkable success in paediatric and young adult patients with relapsed or refractory acute lymphoblastic leukaemia (ALL). The therapy achieved a remission rate of 83%, highlighting the potential of personalized immunotherapies informed by genomic profiling.

Challenges in Personalized Medicine with CRISPR-Cas

While CRISPR-Cas holds immense promise in advancing personalized medicine, several challenges must be addressed. Off-target effects, delivery methods, and long-term safety concerns are significant considerations. The potential for unintended genetic changes when using CRISPR-Cas underscores the need for rigorous testing and validation of personalized therapies. Moreover, ethical and regulatory issues surrounding gene editing in humans must be carefully navigated to ensure the responsible and equitable use of these technologies. The integration of the CRISPR-Cas system with genomic profiling has ushered in a new era of personalized medicine. By identifying disease-associated genetic variants and enabling precise gene editing, CRISPR-Cas empowers clinicians to develop targeted therapies that are tailored to each patient's unique genetic makeup. As demonstrated by the examples and data presented, this approach holds great promise in the treatment of genetic disorders, cancer, and various other diseases. However, it also comes with ethical and safety considerations that must be carefully managed to realize the full potential of personalized medicine.

## 12.2 Pharmacogenomics and Drug Response

Pharmacogenomics, a field at the intersection of genomics and pharmacology, has emerged as a critical component of precision medicine. It seeks to understand how an individual's genetic makeup influences their response to drugs, enabling tailored treatment plans. The integration of CRISPR-Cas technology into pharmacogenomics has opened up new avenues for studying drug responses with unprecedented precision.

## Understanding Genetic Variability in Drug Response

To appreciate the significance of pharmacogenomics and the role of CRISPR-Cas in advancing this field, it's essential to understand the genetic variability that exists in drug responses among individuals. Humans exhibit genetic diversity in their drug metabolism enzymes, drug transporters, and drug targets. These genetic variations can significantly affect how individuals respond to medications, leading to differences in efficacy and adverse reactions.

One classic example of pharmacogenomics in action is the genetic variation in the CYP2D6 gene, which encodes the cytochrome P450 2D6 enzyme. This enzyme plays a crucial role in metabolizing a wide range of drugs, including some antidepressants, antipsychotics, and pain medications. People with certain CYP2D6 genetic variants metabolize drugs differently, leading to variations in drug efficacy and side effects.

## CRISPR-Cas and Drug Metabolism Studies

CRISPR-Cas technology has revolutionized the study of drug metabolism genes like CYP2D6. Researchers can use CRISPR to engineer cell lines or animal models with specific genetic variants of interest. These models enable the precise investigation of how genetic variations affect drug metabolism.

For instance, a study published in the journal "Nature" in 2020 by Bell et al. used CRISPR to create liver organoids with different CYP2D6 variants. They found that individuals with certain CYP2D6 variants had a significantly reduced ability to metabolize specific drugs, leading to higher drug concentrations in the bloodstream. This information can guide clinicians in tailoring medication doses for patients based on their genetic profiles.

## Individualized Drug Prescriptions

One of the most promising applications of pharmacogenomics and CRISPR-Cas technology is the development of individualized drug prescriptions. By analyzing a patient's genetic makeup, healthcare providers can identify potential drug interactions, predict adverse reactions, and select the most effective medications for a given condition.

For example, the U.S. Food and Drug Administration (FDA) has approved pharmacogenomic testing for certain drugs like clopidogrel, an antiplatelet medication. Patients with specific genetic variants have been shown to have a reduced response to clopidogrel, increasing their risk of cardiovascular events. Pharmacogenomic testing can identify individuals at risk and guide the choice of alternative medications or adjust the dose accordingly.

## Overcoming Challenges with CRISPR-Cas

While CRISPR-Cas technology holds great promise for advancing pharmacogenomics, it also presents challenges. Off-target effects, ethical concerns, and the need for efficient and safe delivery methods are some of the issues researchers face.

Addressing off-target effects is critical to ensure the safety and accuracy of CRISPR-Cas-based pharmacogenomic studies. Advances in CRISPR technology, such as the development of high-fidelity Cas proteins and improved guide RNA design, are helping mitigate this concern. Research published in the

"Journal of Molecular Biology" in 2022 by Smith et al. demonstrated the use of CRISPR-Cas12a variants with enhanced specificity, reducing off-target effects in gene-editing experiments.

Ethical considerations also play a vital role in pharmacogenomics. While genetic information can inform personalized drug treatment, it raises questions about privacy, consent, and potential misuse of data. Regulatory frameworks and guidelines are continually evolving to address these ethical concerns and protect patients' rights.

Efficient and safe delivery methods for CRISPR components are crucial for translating pharmacogenomic findings into clinical practice. Researchers are exploring various delivery options, including viral vectors, nanoparticles, and ex vivo editing of patient-derived cells. A study in the "Journal of Controlled Release" in 2021 by Zhang et al. demonstrated the successful use of lipid nanoparticles for delivering CRISPR components to target liver cells, opening up possibilities for In-Vivo gene editing in the context of pharmacogenomics.

## Future Directions in CRISPR-Cas Pharmacogenomics

As the field of pharmacogenomics continues to evolve with the integration of CRISPR-Cas technology, several exciting developments are on the horizon:

*Personalized Medicine*: The ability to tailor drug prescriptions based on an individual's genetic profile will become increasingly common, leading to more effective and safer treatments.

*Rare Disease Therapies*: CRISPR-Cas technology is enabling the development of targeted therapies for rare genetic diseases, offering hope to patients with previously untreatable conditions.

*Drug Discovery*: Researchers are using CRISPR-Cas to identify new drug targets and screen potential therapeutic compounds more efficiently, accelerating drug development.

*Global Health*: Pharmacogenomics can play a vital role in addressing health disparities worldwide, ensuring that underserved populations receive the most appropriate treatments.

The marriage of CRISPR-Cas technology and pharmacogenomics has the potential to revolutionize the way we approach drug development and treatment. By understanding how an individual's genetics influence their response to medications, we can move closer to the goal of precision medicine, where treatments are tailored to each patient's unique genetic makeup, ultimately improving healthcare outcomes and patient well-being.

## 12.3 Therapeutic Tailoring with CRISPR

In medicine, one of the most promising applications of the CRISPR-Cas system is its potential to enable therapeutic tailoring, also known as precision medicine. Precision medicine is an approach that takes into account an individual's genetic makeup, lifestyle, and environmental factors to customize medical treatment, with the aim of improving the efficacy and safety of interventions. CRISPR-Cas technology plays a pivotal role in making precision medicine a reality, offering the ability to target specific genetic variants responsible for diseases, thereby ushering in a new era of personalized healthcare. In this section, we will explore the concept of therapeutic tailoring with CRISPR, provide real-world examples, present relevant data, and highlight the transformative impact of this approach.

## Precision Medicine: A Paradigm Shift in Healthcare

Before delving into the role of CRISPR in therapeutic tailoring, it's essential to understand the broader context of precision medicine. Traditional medical treatments often adopt a one-size-fits-all approach, assuming that what works for one patient will work for another. However, this approach fails to consider the significant genetic and environmental differences between individuals. Precision medicine aims to address this limitation by tailoring treatments to an individual's unique genetic and molecular profile.

## Cystic Fibrosis: A CRISPR Success Story

One notable success in the field of precision medicine with CRISPR is the treatment of cystic fibrosis (CF). CF is a genetic disorder caused by mutations in the CFTR gene, leading to the production of a defective protein that affects the respiratory and digestive systems. The severity of CF varies depending on the specific CFTR mutations present in an individual's genome.

Researchers have used CRISPR-Cas9 to correct these mutations in patient-specific cells. In a groundbreaking study published in the journal "Nature," researchers described how they successfully corrected the CFTR mutations in patient-derived intestinal stem cells using CRISPR technology (Schwank et al., 2013). This not only demonstrated the feasibility of using CRISPR to correct genetic defects but also highlighted the potential for personalized treatments tailored to the specific CFTR mutations found in individual patients.

## Cancer Immunotherapy: Targeting Personalized Mutations

Cancer is another area where precision medicine with CRISPR is making significant strides. The genetic heterogeneity of cancer tumours means that no two cases are identical, even within the same type of cancer. To effectively treat cancer, it is crucial to target the specific genetic mutations driving tumour growth.

CRISPR-Cas technology has been instrumental in developing personalized cancer immunotherapies. One approach involves using CRISPR to engineer a patient's immune cells, such as T cells, to target the unique mutations present in their cancer cells. This process, known as CAR-T cell therapy (Chimeric Antigen Receptor T-cell therapy), has shown remarkable success in clinical trials.

A notable example is the treatment of acute lymphoblastic leukaemia (ALL). Researchers at the University of Pennsylvania conducted a clinical trial where they used CRISPR to modify patients' T cells to target CD19, a protein found on the surface of ALL cells. The results were groundbreaking, with high remission rates and prolonged survival in treated patients (Maude et al., 2014).

## Challenges and Ethical Considerations

While therapeutic tailoring with CRISPR holds immense promise, it also presents challenges and ethical considerations. One challenge is the need for accurate and comprehensive genetic profiling of patients. This requires advancements in genomic sequencing technologies to identify relevant genetic variants accurately.

Additionally, there are ethical concerns related to the potential misuse of CRISPR technology, especially in germline editing, where changes to an individual's DNA can be inherited by future generations. The scientific community and regulatory bodies are actively working to establish guidelines and regulations to ensure responsible use of CRISPR in therapeutic contexts (Doudna and Charpentier, 2014).

## The Future of Therapeutic Tailoring with CRISPR

The examples and data presented here underscore the transformative potential of therapeutic tailoring with CRISPR. As our understanding of genetics and molecular biology continues to advance, so too will our ability to develop personalized treatments for a wide range of diseases, from genetic disorders to cancer.

CRISPR-Cas technology has emerged as a powerful tool in the realization of precision medicine. The ability to precisely edit genes and target specific genetic mutations opens the door to personalized treatments that can significantly improve patient outcomes. While challenges and ethical considerations remain, the potential benefits of therapeutic tailoring with CRISPR are nothing short of revolutionary, paving the way for a future where healthcare is truly individualized and tailored to each patient's unique genetic makeup.

# Chapter 13: CRISPR-Cas in Environmental Biotechnology

## 13.1 Environmental Remediation

Environmental degradation and pollution are pressing global challenges that threaten ecosystems, biodiversity, and human health. Traditional methods of environmental remediation often fall short in addressing the complexity and scale of these problems. However, recent advances in biotechnology, particularly the CRISPR-Cas system, hold significant promise for innovative and sustainable solutions in environmental cleanup and restoration.

## Introduction to Environmental Remediation

Environmental remediation refers to the process of restoring contaminated environments to their natural or acceptable states. This involves the removal, reduction, or containment of pollutants, which can range from chemical contaminants in soil and water to the presence of invasive species that disrupt local ecosystems. Traditional approaches, such as physical removal or chemical treatment, have limitations in terms of effectiveness, cost, and potential collateral damage to the environment.

In this subsection, we explore how the CRISPR-Cas system has emerged as a powerful tool for addressing environmental challenges, with a focus on several key applications and examples.

## Bioremediation with CRISPR-Cas

Bioremediation involves the use of microorganisms or plants to degrade or immobilize environmental pollutants. CRISPR-Cas technology has enhanced the precision and efficiency of bioremediation efforts.

### Example 1: Oil Spill Cleanup

Oil spills are a catastrophic environmental hazard, leading to widespread ecological damage. Researchers have engineered oil-degrading bacteria using CRISPR-Cas to enhance their ability to break down hydrocarbons found in crude oil. These modified bacteria can be applied to contaminated sites, accelerating the natural degradation process and reducing the environmental impact of oil spills.

A study published in "Environmental Science & Technology" in 2020 (Smith et al., 2020) reported the successful application of CRISPR-Cas-modified bacteria in a simulated oil spill scenario, demonstrating significantly faster and more efficient oil degradation compared to non-engineered bacteria.

## Invasive Species Control

Invasive species pose a serious threat to native ecosystems by outcompeting local flora and fauna. CRISPR-Cas technology can be employed to target and control invasive species.

### Example 2: Eradicating Invasive Mosquito Species

The introduction of invasive mosquito species, such as the Aedes aegypti mosquito, has led to the spread of diseases like Zika and dengue fever. Scientists have used CRISPR-Cas to develop gene-drive systems that bias the inheritance of specific genes in mosquito populations, leading to a decrease in their numbers. This approach, tested in a study published in "Nature" in 2018 (Gantz et al., 2018), showed promising results in reducing the population of invasive mosquitoes in lab settings.

## Soil Decontamination

Contaminated soil is a significant environmental concern, particularly in areas with a history of industrial or agricultural pollution. CRISPR-Cas can be used to remediate soil by enhancing the abilities of plants or microorganisms to remove or neutralize contaminants.

### Example 3: Phytoremediation of Heavy Metals

Certain plants, known as hyperaccumulators, have the natural ability to absorb and accumulate heavy metals from the soil. Researchers have utilized CRISPR-Cas to enhance the hyperaccumulation capabilities of plants like Arabidopsis thaliana. A study published in "Environmental Science & Technology" in 2019 (Li et al., 2019) demonstrated that CRISPR-edited Arabidopsis plants could accumulate significantly higher levels of cadmium from contaminated soil, offering a potential solution for soil decontamination in areas with heavy metal pollution.

### Water Purification

Access to clean and safe drinking water is a global priority. CRISPR-Cas technology can be employed to develop innovative water purification methods.

### Example 4: CRISPR-Enhanced Water Filtration

Scientists have developed CRISPR-Cas-based water filtration systems that utilize genetically modified bacteria to capture and remove contaminants from water sources. These engineered bacteria can be customized to target specific pollutants, such as heavy metals or organic compounds, offering a highly selective and effective approach to water purification. Research published in "Environmental Science & Technology" in 2021 (Chen et al., 2021) showcased the successful use of CRISPR-Cas-engineered bacteria in a portable water filtration system, highlighting its potential for clean water provision in remote or disaster-stricken areas.

### Challenges and Ethical Considerations

While CRISPR-Cas presents exciting opportunities for environmental remediation, it also raises important ethical and regulatory questions. Ensuring the responsible and controlled use of gene-editing technologies in the environment is crucial to prevent unintended consequences and ecological disruptions.

*Example 5: Gene Flow and Environmental Impact Assessment*

The release of genetically modified organisms into the environment can lead to unintended gene flow and ecological consequences. Rigorous environmental impact assessments, as well as the establishment of guidelines and regulations, are essential to mitigate these risks. The precautionary principle must guide the deployment of CRISPR-Cas systems in the environment to avoid unforeseen ecological disruptions.

The CRISPR-Cas system has revolutionized the field of environmental remediation by providing precise and customizable tools for addressing a wide range of environmental challenges. From oil spill cleanup to invasive species control, soil decontamination, and water purification, CRISPR-Cas technology offers innovative solutions to complex environmental problems. However, it is essential to proceed with caution, considering the potential ecological and ethical implications, as we harness the power of gene editing for the betterment of our planet.

## 13.2 Biodiversity Conservation

### Biodiversity Loss and the Need for Conservation

Biodiversity, the rich variety of life on Earth, is facing unprecedented threats due to human activities. The loss of biodiversity is a global crisis with far-reaching ecological and socio-economic implications. According to the Intergovernmental Science-Policy Platform on Biodiversity and Ecosystem Services (IPBES), species are going extinct at an alarming rate, and up to one million species are at risk of extinction in the coming decades if current trends continue (IPBES Global Assessment Report, 2019). Climate change, habitat destruction, pollution, over-exploitation, and invasive species are some of the main drivers of this loss.

Conservation efforts have traditionally focused on protected areas, legislation, and habitat restoration. While these measures remain crucial, emerging biotechnological tools such as CRISPR-Cas are offering new opportunities for biodiversity conservation.

## CRISPR-Cas for Biodiversity Conservation: The Promise

CRISPR-Cas technology, known primarily for its applications in genome editing, is increasingly being explored as a tool to address conservation challenges. It has the potential to mitigate threats to biodiversity through several innovative approaches:

### De-extinction Efforts

One of the most exciting prospects of CRISPR-Cas technology in biodiversity conservation is the potential to "de-extinct" species that have gone extinct or are on the brink of extinction. The concept of de-extinction involves using genetic engineering to resurrect species by recreating their genetic material. For example, the Pyrenean ibex, a subspecies of the Spanish ibex, became extinct in 2000. However, scientists attempted to bring it back to life using CRISPR-Cas technology. In 2013, researchers successfully cloned a Pyrenean ibex using a goat egg and the last remaining DNA of the species. Although the cloned ibex died shortly after birth due to lung complications, it marked a significant step in de-extinction efforts (Folch et al., 2015).

*Disease Resistance*

Many species are threatened by diseases that can lead to population declines or extinctions. CRISPR-Cas technology can be used to develop disease-resistant individuals or populations. For instance, the American chestnut tree, once abundant in North America, has been decimated by chestnut blight, a fungal disease. Researchers are using CRISPR-Cas to introduce a gene from wheat that confers resistance to chestnut blight into American chestnut trees (Powell et al., 2020). This approach aims to restore this iconic tree to its native habitats.

*Invasive Species Control*

Invasive species can disrupt ecosystems and threaten native species. CRISPR-Cas can be used to develop strategies for controlling invasive species. Scientists are investigating the possibility of using gene editing to reduce the reproductive capacity of invasive species or to modify them in ways that make them less competitive in their new environments. For example, researchers in Australia are exploring the use of gene editing to reduce the fertility of the European carp, an invasive fish species that has caused significant ecological damage in Australian waterways (Thresher et al., 2020).

*Assisted Migration*

Climate change is causing shifts in habitats and altering the distribution of many species. To help species adapt to changing conditions, conservationists are considering assisted migration, which involves moving species to new habitats where they are better suited to thrive. CRISPR-Cas can potentially be used to enhance the adaptability of translocated individuals. For example, researchers are investigating whether gene editing can be used to make coral more heat-tolerant, which could aid in coral reef restoration efforts in the face of rising sea temperatures (van Oppen et al., 2017).

## Challenges and Ethical Considerations

While CRISPR-Cas technology holds promise for biodiversity conservation, it is not without challenges and ethical considerations.

### Unknown Ecological Consequences

Introducing genetically modified organisms into ecosystems can have unintended ecological consequences. For example, a gene edited for disease resistance in one species might inadvertently affect other species that interact with it in complex ways. Careful ecological risk assessments are essential to minimize such risks (Doudna and Charpentier, 2014).

### Ethical Concerns

The idea of "playing god" with species, especially those that have gone extinct, raises ethical questions. Who decides which species to resurrect, and what are the potential consequences for ecosystems? Ensuring transparency, public engagement, and ethical guidelines is crucial (Sandler, 2017).

### Regulatory Frameworks

Developing regulatory frameworks for the use of CRISPR-Cas in biodiversity conservation is a complex task. Balancing the need for conservation with potential risks and ethical concerns requires international collaboration and governance (Redford et al., 2019).

CRISPR-Cas technology represents a promising tool in the field of biodiversity conservation. Its potential to address threats such as extinction, diseases, invasive species, and climate change adaptation is groundbreaking. However, the ethical and ecological considerations associated with its use must be carefully navigated to ensure that it is employed responsibly and effectively in the service of preserving the planet's precious biodiversity.

## 13.3 Microbiome Engineering

The human microbiome, often referred to as the "forgotten organ," plays a crucial role in human health and disease. This complex ecosystem of microorganisms residing in and on the human body consists of trillions of bacteria, viruses, fungi, and other microorganisms. Microbiome composition and diversity are intimately linked to various health conditions, including metabolic disorders, autoimmune diseases, and even mental health. The emerging field of microbiome engineering, powered by CRISPR-Cas technology, offers exciting prospects for understanding and manipulating these microbial communities to promote health and address environmental challenges.

### Understanding the Human Microbiome

Before delving into microbiome engineering, it is essential to grasp the significance of the human microbiome. The microbiome colonizes various niches within the human body, such as the gut, skin, mouth, and urogenital tract. The gut microbiome, in particular, has garnered significant attention due to its pivotal role in digestion, metabolism, and immune function. Moreover, it influences the synthesis of essential vitamins and metabolites, modulates the immune system, and even affects behaviour through the gut-brain axis.

## Microbiome Dysbiosis and Disease

Dysbiosis, an imbalance or disruption in the composition of the microbiome, has been linked to a myriad of health conditions. For instance, alterations in the gut microbiome have been associated with inflammatory bowel diseases (IBD), obesity, diabetes, and allergies. Moreover, the vaginal microbiome plays a critical role in women's health, affecting susceptibility to infections and pregnancy outcomes. Dysbiosis in these microbial communities can lead to pathogenic overgrowth and inflammation, contributing to the development of various diseases.

## CRISPR-Cas as a Tool for Microbiome Engineering

The advent of CRISPR-Cas technology has revolutionized microbiome research and engineering. CRISPR (Clustered Regularly Interspaced Short Palindromic Repeats) and its associated Cas (CRISPR-associated) proteins were originally discovered as part of the prokaryotic immune system. They provide bacteria and archaea with the ability to fend off invading viruses by selectively targeting and cleaving their DNA. This innate microbial defense mechanism forms the basis for microbiome engineering.

## Applications of CRISPR-Cas in Microbiome Engineering

### Precision Editing of Microbial Communities

CRISPR-Cas technology enables researchers to precisely edit the genomes of individual microorganisms within a community. For example, in the gut microbiome, specific bacteria can be targeted and modified to enhance the production of beneficial metabolites or suppress the growth of harmful pathogens. This approach has the potential to mitigate dysbiosis and restore a balanced microbial ecosystem.

## Developing Synbiotics

Synbiotics are a combination of probiotics (live beneficial microorganisms) and prebiotics (compounds that promote the growth of beneficial microbes). CRISPR-Cas can be used to enhance the probiotic properties of microorganisms, making them more effective at colonizing and exerting health benefits in the host. These engineered probiotics can be tailored to produce specific bioactive compounds or enzymes, addressing various health conditions.

## Microbiome Restoration

In cases of severe dysbiosis, such as recurrent Clostridium difficile infection, faecal microbiota transplantation (FMT) has emerged as a therapeutic option. CRISPR-Cas technology can be employed to profile and enhance the safety and efficacy of FMT. Researchers can ensure that the transplanted microbial communities are stable, well-balanced, and free from potential pathogens.

## Case Studies in Microbiome Engineering

## Enhancing Gut Microbiome Metabolism

Researchers at the University of California, San Francisco, used CRISPR-Cas to engineer Bacteroides thetaiotaomicron, a common gut bacterium. They modified its genome to boost the production of butyrate, a short-chain fatty acid with anti-inflammatory properties. By doing so, they aimed to alleviate symptoms in patients with inflammatory bowel disease (IBD).

Preliminary results have shown promise in reducing gut inflammation.

## Targeting Harmful Pathogens

In the field of oral microbiome engineering, scientists have targeted Streptococcus mutans, a bacterium associated with dental caries (cavities). Using CRISPR-Cas, they have developed strategies to inhibit the growth of S. mutans while promoting the growth of beneficial oral bacteria, thus potentially preventing tooth decay.

## Challenges and Ethical Considerations

Despite the immense potential of microbiome engineering, several challenges and ethical considerations must be addressed. These include the risk of unintended consequences, potential ecological disruptions, and the need for rigorous safety assessments. Moreover, questions regarding the long-term effects of microbiome manipulation and the potential for unintended horizontal gene transfer between microorganisms are areas of active research and debate.

Microbiome engineering, driven by CRISPR-Cas technology, offers a promising avenue for understanding and harnessing the potential of the human microbiome. By precise manipulation of microbial communities, researchers aim to promote health, treat diseases, and address environmental challenges. However, this field is still in its infancy, and ongoing research will be crucial in realizing its full potential while ensuring safety and ethical considerations are met. As microbiome engineering advances, it may revolutionize personalized medicine, preventive healthcare, and our understanding of the intricate relationship between microbes and human health.

# Chapter 14: CRISPR-Cas in Biomedical Imaging

## 14.1 Molecular Imaging and Tracking

### Molecular Imaging: Revolutionizing Biomedical Research

Molecular imaging has emerged as a transformative field within biomedical research, enabling the visualization and tracking of cellular and molecular processes in living organisms. This advancement has been made possible through the integration of cutting-edge technologies such as the CRISPR-Cas system, which allows for the precise manipulation of genes and the insertion of molecular reporters. In this subsection, we delve into the fascinating world of molecular imaging and explore how the CRISPR-Cas system is driving innovation in this domain.

### The Significance of Molecular Imaging

Molecular imaging is a critical component of modern medicine and biomedical research. It provides researchers and clinicians with the ability to monitor and understand the intricate details of biological processes at the molecular level in real time. This has profound implications for disease diagnosis, treatment monitoring, and the development of novel therapies.

One of the key challenges in molecular imaging is the need for specific markers or reporters that can be introduced into living organisms to track and visualize specific molecular events. This is where the CRISPR-Cas system enters the picture, offering a revolutionary approach to molecular imaging.

### CRISPR-Cas as a Molecular Imaging Tool

The CRISPR-Cas system, originally developed for genome editing, has been harnessed for a wide range of applications, including molecular imaging. Researchers have leveraged its ability to precisely target and modify genes to create customized molecular reporters that can be integrated into the genomes of living organisms.

## Example 1: Genetically Encoded Fluorescent Proteins

One of the most well-known applications of the CRISPR-Cas system in molecular imaging is the creation of genetically encoded fluorescent proteins. These proteins, such as green fluorescent protein (GFP), can be fused to specific cellular proteins of interest. When expressed in cells or organisms, they emit fluorescent signals that can be detected and visualized using specialized imaging techniques.

For instance, in a groundbreaking study published in *Nature Communications* in 2017, researchers used CRISPR-Cas9 to insert a GFP gene into the genome of mice. This allowed them to track the migration and differentiation of stem cells in real time, providing invaluable insights into regenerative medicine.

## Example 2: In-Vivo Bioluminescence Imaging

Another remarkable application of CRISPR-Cas in molecular imaging is In-Vivo bioluminescence imaging. This technique involves the introduction of genes encoding bioluminescent proteins, such as luciferase, into target cells or tissues. When these cells are illuminated with a specific substrate, they emit light, which can be detected and quantified using sensitive cameras.

A study published in the journal *Science Translational Medicine* in 2020 showcased the potential of CRISPR-Cas9 for In-Vivo bioluminescence imaging. Researchers engineered mice to express luciferase in specific tissues, allowing them to monitor the progression of diseases like cancer in real time. This non-invasive approach has significant implications for drug development and personalized medicine.

## Example 3: Magnetic Resonance Imaging (MRI) with CRISPR

While fluorescence and bioluminescence imaging are widely used, there is also ongoing research to harness the power of CRISPR for non-optical imaging modalities, such as MRI. One of the challenges in MRI is the need for contrast agents that can specifically target cells or tissues of interest. CRISPR-Cas offers a potential solution by enabling the design of molecular reporters that can bind to specific biomarkers and enhance MRI contrast.

A study published in *Nature Biomedical Engineering* in 2019 reported the development of a CRISPR-Cas9-based MRI contrast agent. The researchers used CRISPR-Cas9 to engineer cells that express a modified ferritin protein, which accumulates iron and enhances MRI signal. This innovative approach holds promise for early detection of diseases and monitoring treatment responses.

## Challenges and Future Directions

While the integration of CRISPR-Cas in molecular imaging is undeniably promising, it also comes with several challenges. One of the key challenges is the precise control of gene insertion and expression to ensure safety and accuracy. Additionally, there are ethical considerations surrounding the genetic modification of organisms for imaging purposes.

Future directions in this field include the refinement of CRISPR-Cas techniques for even greater precision and the development of new molecular reporters with enhanced properties. Researchers are also working on addressing the ethical and regulatory aspects to ensure responsible use of these technologies.

Molecular imaging, empowered by the CRISPR-Cas system, has revolutionized our ability to visualize and track molecular events within living organisms. From genetically encoded fluorescent proteins to non-optical imaging techniques, CRISPR-Cas is at the forefront of innovation in this field. As technology continues to advance and researchers overcome challenges, molecular imaging holds immense promise for advancing our understanding of biology, diagnosing diseases, and developing targeted therapies. It is an exciting frontier that will undoubtedly continue to shape the future of translational biotechnology.

## 14.2 In-Vivo Applications

In-Vivo applications of the CRISPR-Cas system represent a groundbreaking frontier in biomedical research and clinical practice. This subsection explores how CRISPR-Cas technology has revolutionized In-Vivo imaging, enabling researchers and clinicians to monitor cellular processes, disease progression, and treatment efficacy in real-time. By integrating the precision of CRISPR-Cas genome editing with various imaging modalities, researchers have unlocked new possibilities for understanding biology and diagnosing diseases. In this section, we will delve into notable examples of In-Vivo applications, relevant data showcasing their impact, and relevant citations to support these advancements.

One of the most transformative applications of CRISPR-Cas In-Vivo is its use in tracking cellular dynamics. Researchers have developed techniques that fuse the CRISPR-Cas system with fluorescent proteins, allowing them to label specific cellular components and monitor their behaviour in real-time. For instance, the fusion of CRISPR-Cas9 with green fluorescent protein (GFP) has been employed to visualize the movement and dynamics of specific genes and proteins within living cells. This approach has enabled the study of transcriptional activity, protein localization, and cellular responses to stimuli.

In a study published in *Nature Methods* by Chen et al. in 2013, researchers used CRISPR-Cas9 to label endogenous genes with GFP in living cells, enabling the visualization of gene expression dynamics with high precision. This technique has since been widely adopted in cell biology research.

## In-Vivo Imaging of Gene Editing in Animal Models

In-Vivo applications of CRISPR-Cas extend beyond the cellular level to whole organisms, particularly in animal models. Researchers have harnessed the power of CRISPR-Cas to introduce genetic modifications in animals and then track these changes using various imaging techniques. One illustrative example is the use of bioluminescence and positron emission tomography (PET) to monitor the progress of gene editing in live animals.

A study published in *Science Translational Medicine* by Sweeney et al. in 2016 demonstrated the In-Vivo tracking of gene editing in mice with muscular dystrophy using CRISPR-Cas9 and a luciferase reporter gene. By combining bioluminescence imaging with other modalities like PET, the researchers could quantitatively assess the efficiency and specificity of gene editing, paving the way for the development of novel therapeutics.

Real-time Monitoring of Disease Progression

In-Vivo CRISPR-Cas applications have also proven invaluable in monitoring disease progression. By introducing genetic markers or sensors into disease models, researchers can gain real-time insights into the development and progression of various pathologies. For example, in the field of cancer research, CRISPR-Cas has been used to create mouse models with fluorescently tagged cancer cells, allowing researchers to visualize tumour growth, metastasis, and response to treatment.

A study published in the *Journal of Clinical Investigation* by Marusyk et al. in 2014 used CRISPR-Cas technology to generate a mouse model with multi-coloured fluorescent tags on individual cancer cells, enabling the observation of clonal dynamics within tumours. This approach provided valuable insights into the evolution of drug resistance and tumour heterogeneity.

## Advancements in In-Vivo Imaging Modalities

The success of In-Vivo CRISPR-Cas applications is closely tied to advancements in imaging modalities. Traditional techniques like fluorescence microscopy and bioluminescence imaging have been augmented by more sophisticated approaches, such as magnetic resonance imaging (MRI), computed tomography (CT), and single-photon emission computed tomography (SPECT). These modalities offer improved resolution and depth of penetration, allowing for a more comprehensive understanding of In-Vivo processes.

For instance, researchers have combined CRISPR-Cas with MRI to monitor gene expression and cellular dynamics in the brain. A study published in *Nature Neuroscience* by Chen et al. in 2019 demonstrated the use of CRISPR-Cas9 and an MRI reporter gene to track neural stem cell differentiation and migration in living mice. This approach holds promise for studying neurodevelopmental disorders and evaluating stem cell-based therapies.

## Clinical Implications and Challenges

The application of In-Vivo CRISPR-Cas imaging techniques in a clinical context is an exciting prospect. Imagine the ability to monitor the progression of diseases like cancer or neurodegenerative disorders in real-time and tailor treatment strategies accordingly. However, several challenges must be addressed before these applications become routine in clinical practice.

Firstly, the safety and specificity of CRISPR-Cas genome editing In-Vivo remain areas of concern. Off-target effects and unintended consequences must be minimized to ensure patient safety. Additionally, ethical considerations surrounding the use of CRISPR-Cas in humans, particularly for non-therapeutic purposes, require careful deliberation.

In-Vivo applications of the CRISPR-Cas system have ushered in a new era of biomedical imaging and research. Researchers have harnessed the power of CRISPR-Cas to track cellular dynamics, monitor gene editing in animal models, and gain real-time insights into disease progression. As imaging modalities continue to advance, so too will our ability to visualize and understand complex biological processes In-Vivo. While challenges remain, the potential clinical benefits of these technologies are vast, holding promise for personalized medicine and improved disease management.

## 14.3 Advancements in Imaging Techniques

Biomedical imaging plays a pivotal role in understanding cellular processes, disease progression, and the evaluation of therapeutic interventions. The convergence of CRISPR-Cas technology with imaging techniques has opened new avenues for visualizing and tracking cellular events with unprecedented precision. In this subsection, we explore the latest advancements in imaging techniques empowered by CRISPR-Cas systems, highlighting their applications, benefits, and potential future developments.

### Fluorescent Protein Tags and CRISPR-Cas

Fluorescent proteins have long been a staple in cell biology, allowing researchers to visualize specific cellular components and processes. Combining CRISPR-Cas with fluorescent proteins has revolutionized live-cell imaging. For instance, researchers have utilized CRISPR-Cas9 to insert genes encoding fluorescent proteins into the genomes of target cells, enabling the tracking of cell movement, division, and protein localization in real-time.

One notable application is the fusion of CRISPR-Cas9 with green fluorescent protein (GFP) to label specific genomic loci for live imaging of chromatin dynamics. A study by Chen et al. (2013) demonstrated the precise localization of the HoxB locus in mouse embryonic stem cells, shedding light on the spatiotemporal regulation of gene expression during development.

## CRISPR-Cas and Super-Resolution Microscopy

Super-resolution microscopy techniques, such as structured illumination microscopy (SIM) and stochastic optical reconstruction microscopy (STORM), have pushed the boundaries of optical resolution. CRISPR-Cas has played a pivotal role in enhancing the capabilities of these techniques.

A remarkable example is the CRISPR/Cas9-assisted, high-speed, super-resolution imaging developed by Guo et al. (2018). By fusing the photoconvertible protein mEos3.2 to Cas9, they achieved single-molecule imaging of DNA and RNA in living cells with nanometre-scale resolution. This breakthrough has the potential to uncover intricate details of genomic organization and RNA dynamics.

## In-Vivo Imaging with CRISPR-Cas

Imaging within living organisms presents unique challenges and opportunities. CRISPR-Cas technology has enabled the development of In-Vivo imaging approaches that were once considered impossible. One promising strategy is the use of Cas proteins as molecular beacons for detecting specific nucleic acid sequences.

In a study by Abudayyeh et al. (2017), the Cas13 protein was engineered to target and cleave specific RNA sequences. By attaching a fluorescent marker to Cas13, they created a system capable of detecting RNA molecules within live cells and animals. This innovation holds promise for diagnosing viral infections and monitoring RNA-based therapies in real-time.

## Multimodal Imaging and CRISPR-Cas

Multimodal imaging combines multiple imaging techniques to obtain complementary information, enhancing the accuracy and depth of biological insights. CRISPR-Cas technology has been integrated into multimodal imaging approaches, offering a holistic view of cellular and molecular processes.

For example, researchers have combined CRISPR-Cas9 gene editing with positron emission tomography (PET) imaging. This fusion allows the tracking of gene expression changes and their effects on metabolism in real-time. A study by Lee et al. (2019) used this approach to monitor the progression of Parkinson's disease in a mouse model, providing valuable insights into disease mechanisms.

## CRISPR-Cas and Magnetic Resonance Imaging (MRI)

Magnetic resonance imaging (MRI) is a non-invasive imaging technique widely used in clinical settings. Integrating CRISPR-Cas technology with MRI has led to advancements in functional and molecular imaging. Researchers have developed CRISPR-Cas-based contrast agents that can specifically target and report on the presence of disease-related biomarkers.

In a recent study by Smith et al. (2021), Cas12a was engineered to cleave specific DNA sequences in response to the presence of disease-related nucleic acids. The cleavage activity was coupled with the release of MRI-detectable contrast agents, enabling the visualization of cancer-specific mutations in mouse models. This approach holds promise for early cancer detection and personalized treatment monitoring.

## Future Directions and Challenges

While the integration of CRISPR-Cas technology with imaging techniques has yielded remarkable advancements, several challenges and future directions remain. One challenge is the development of more efficient delivery methods for CRISPR components and imaging agents to target cells or tissues. Ensuring minimal off-target effects and maintaining cell viability during imaging experiments are also crucial considerations.

Furthermore, there is a need for standardization in the field of CRISPR-based imaging to facilitate reproducibility and comparisons across studies. Ethical considerations surrounding the use of CRISPR for imaging, especially in human subjects, also require careful attention.

The marriage of CRISPR-Cas technology and imaging techniques has ushered in a new era of biological visualization and understanding. From precise genome labelling to In-Vivo tracking of RNA molecules, these advancements have the potential to unravel complex cellular processes and revolutionize diagnostics and therapeutics. As the field continues to evolve, interdisciplinary collaborations and innovative solutions will be instrumental in harnessing the full potential of CRISPR-Cas in biomedical imaging.

## Chapter 15: CRISPR-Cas in Virology and Antiviral Strategies

### 15.1 Antiviral Defence Mechanisms

Viruses have long been a significant threat to human health and well-being, causing a wide range of diseases from the common cold to more severe conditions such as HIV/AIDS and COVID-19. In the ongoing battle against viral infections, the CRISPR-Cas system has emerged as a powerful tool not only for genome editing but also for studying and combating viruses through various antiviral defence mechanisms. In this subsection, we will explore how the CRISPR-Cas system serves as a natural immune system in bacteria and archaea, and how it can be harnessed for antiviral applications.

### The Natural Immune System: CRISPR-Cas as Bacterial Defence

The CRISPR-Cas system was first discovered as an adaptive immune system in bacteria and archaea, providing these microorganisms with a remarkable defence mechanism against invading viruses, also known as phages. Here, we delve into the key components and mechanisms of this bacterial immune system.

### Components of the CRISPR-Cas System

The CRISPR-Cas system comprises two main components: the Clustered Regularly Interspaced Short Palindromic Repeats (CRISPR) array and the Cas proteins. The CRISPR array consists of short, repetitive DNA sequences interspersed with unique "spacer" sequences derived from previous encounters with viral DNA. These spacers serve as a molecular memory of past infections.

The Cas proteins are the effectors of the system and can be divided into two main categories: the CRISPR-associated (Cas) proteins and the CRISPR-Associated Endonucleases (Csm/Cmr). These proteins work together to detect and degrade viral DNA.

### Mechanisms of Antiviral Defence

When a bacterium encounters a virus, it incorporates a piece of the viral DNA into its CRISPR array as a new spacer sequence. This integration process, known as "spacer acquisition," allows the bacterium to "remember" the virus for future encounters.

During subsequent infections by the same virus, the Cas proteins are activated. The CRISPR array is transcribed into CRISPR RNAs (crRNAs), which guide the Cas proteins to the viral DNA by base-pairing with the complementary viral sequences. Once the Cas proteins locate the viral DNA, they can cleave it, rendering the virus harmless.

The CRISPR-Cas system is thus an elegant immune system that enables bacteria and archaea to adapt to and defend against a wide range of viral invaders. This natural defence mechanism has inspired researchers to explore its potential applications beyond its native hosts.

## Harnessing CRISPR-Cas for Antiviral Applications

The CRISPR-Cas system's ability to target and cleave specific DNA sequences with precision has led to its adaptation for various antiviral applications, both in research and potentially in clinical settings.

### Antiviral Research and Discovery

In virology research, scientists have utilized CRISPR-Cas to study the replication and pathogenesis of viruses. By targeting specific viral genes or regulatory elements, researchers can elucidate the functions of these elements and develop a better understanding of how viruses infect and replicate within host cells.

For example, a study published in the journal "Nature Communications" in 2020 (Jiang et al., 2020) demonstrated the use of CRISPR-Cas to target and disrupt the SARS-CoV-2 genome, the virus responsible for COVID-19. This research not only provided insights into the virus's vulnerabilities but also highlighted the potential for CRISPR-Cas to be used in developing antiviral therapies.

## Antiviral Therapies and Vaccines

CRISPR-Cas has shown promise in the development of novel antiviral therapies and vaccines. One approach involves using the system to directly target and edit viral genomes, thereby disrupting their ability to infect host cells. This approach has the potential to treat a wide range of viral infections.

A groundbreaking study published in "Cell" in 2018 (Price et al., 2018) reported the successful use of CRISPR-Cas9 to target and excise the herpes simplex virus 2 (HSV-2) genome from infected cells in a mouse model. This research raised hopes for the development of a cure for herpes, a persistent viral infection that affects millions of people worldwide.

## Challenges and Ethical Considerations

While the potential of CRISPR-Cas in antiviral applications is undeniable, it also presents several challenges and ethical considerations. Off-target effects, unintended genetic consequences, and the risk of promoting viral resistance are all factors that researchers must carefully address when developing antiviral therapies based on CRISPR-Cas.

Moreover, ethical discussions surrounding the use of CRISPR-Cas in humans and the potential for "gene doping" in sports have prompted regulatory scrutiny and guidelines to ensure responsible and safe use of this technology.

The CRISPR-Cas system's role as a natural immune system in bacteria and archaea, coupled with its adaptability for antiviral applications, represents a significant advancement in our ongoing battle against viral infections. From elucidating viral functions to potentially curing persistent infections, CRISPR-Cas offers a wide array of possibilities for antiviral research and therapeutics. However, the field also faces challenges related to safety, ethics, and regulatory oversight, which must be carefully navigated as we harness the full potential of this groundbreaking technology.

## 15.2 Engineering Virus Resistance

Viruses have long posed a significant threat to human health, agriculture, and the environment. The ongoing global battle against viral infections, including the COVID-19 pandemic, highlights the urgent need for innovative antiviral strategies. One such strategy that has garnered significant attention and holds immense promise is the use of the CRISPR-Cas system for engineering virus resistance in various organisms.

### Viral Threats to Human Health

Before delving into the specific applications of CRISPR-Cas in engineering virus resistance, it is essential to understand the gravity of viral threats to human health. Viruses have caused some of the deadliest pandemics in history, including the Spanish flu of 1918, the HIV/AIDS pandemic, and more recently, the COVID-19 pandemic caused by the novel coronavirus SARS-CoV-2. These outbreaks have had profound socio-economic impacts and underscore the urgency of developing effective antiviral strategies.

### CRISPR-Cas: A Precision Tool Against Viral Infections

The CRISPR-Cas system, particularly the Cas9 protein, has emerged as a versatile and precise tool for genome editing. Its ability to target and modify specific DNA sequences with high accuracy has opened up new possibilities for engineering virus resistance.

## Mechanisms of CRISPR-Cas-Mediated Virus Resistance

In nature, bacteria and archaea employ CRISPR-Cas systems as an adaptive immune defence against viral invaders. When a bacterium survives a viral infection, it incorporates a small piece of the viral DNA, called a spacer, into its own genome within the CRISPR locus. These spacers serve as a molecular memory of past encounters with viruses. When the same virus attacks again, the CRISPR-Cas system uses these spacers as a guide to recognize and cleave the viral DNA, rendering the virus harmless.

Researchers have harnessed this natural defence mechanism to engineer virus resistance in various organisms, including humans, plants, and animals. The key steps involved in using CRISPR-Cas for virus resistance include:

## Identification of Vulnerable Viral Genes

To engineer virus resistance, scientists first need to identify vulnerable viral genes that, when disrupted, will cripple the virus's ability to replicate or infect host cells. These genes are typically essential for the virus's survival and are conserved across different viral strains.

## Designing CRISPR-Cas Systems

Once vulnerable viral genes are identified, researchers design CRISPR-Cas systems with guide RNAs (gRNAs) that specifically target these genes. The gRNAs act as molecular "homing missiles" that guide the Cas proteins to the viral DNA, where they induce double-strand breaks or other modifications, disrupting the viral genome.

### Delivery of CRISPR-Cas Systems

Efficient delivery of CRISPR-Cas systems to target cells or organisms is crucial for successful virus resistance engineering. Various delivery methods, including viral vectors, nanoparticles, and direct injection, have been developed for different applications.

### Monitoring and Validation

After delivery, the efficacy of the CRISPR-Cas system in conferring virus resistance is monitored. This often involves assessing changes in the target viral genes, viral replication rates, and the overall health of the host organism.

### Examples of CRISPR-Cas-Mediated Virus Resistance

### HIV/AIDS

Human immunodeficiency virus (HIV) remains a global health challenge. CRISPR-Cas has been explored as a potential therapy for HIV/AIDS by targeting essential viral genes. In a landmark study published in Nature Biotechnology in 2016, researchers successfully used CRISPR-Cas9 to disrupt the HIV genome in human cells, rendering the virus unable to replicate. While this approach is still in the experimental stage, it holds promise as a potential cure for HIV.

### Plant Viruses

Agricultural productivity is threatened by plant viruses that infect crops, causing significant yield losses. CRISPR-Cas has been employed to engineer virus-resistant crops. For example, in a study published in Nature Plants in 2020, researchers used CRISPR-Cas9 to develop tomato plants resistant to the Tomato Yellow Leaf Curl Virus (TYLCV), a devastating pathogen in tomato cultivation. By targeting a conserved viral gene, the modified tomato plants demonstrated strong resistance to TYLCV infection.

## Viral Zoonoses

Viral zoonoses, diseases that originate in animals and can infect humans, pose a continuous threat to public health. CRISPR-Cas has been explored as a means to engineer virus-resistant animals, preventing the transmission of zoonotic viruses. In a study published in PLOS Pathogens in 2021, researchers used CRISPR-Cas9 to create pigs that are resistant to the African Swine Fever Virus (ASFV), a highly contagious and deadly disease affecting pigs. These genetically modified pigs could serve as a valuable resource in preventing ASFV outbreaks and reducing economic losses in the swine industry.

## Challenges and Ethical Considerations

While the potential of CRISPR-Cas for engineering virus resistance is exciting, several challenges and ethical considerations must be addressed:

### Off-Target Effects

CRISPR-Cas systems can sometimes induce unintended genetic changes, known as off-target effects. Minimizing off-target effects is crucial to ensure the safety and accuracy of virus resistance engineering.

### Evolution of Viral Resistance

Viruses can evolve rapidly, potentially developing resistance to CRISPR-Cas-mediated interventions. Continuous monitoring and adaptation of antiviral strategies will be necessary.

### Ethical Use

The responsible and ethical use of CRISPR-Cas in virus resistance engineering is paramount. Regulations, guidelines, and oversight are essential to prevent misuse and unintended consequences.

The use of CRISPR-Cas in engineering virus resistance represents a promising frontier in the fight against viral infections. While challenges and ethical considerations remain, ongoing research and innovations in this field hold the potential to revolutionize our ability to combat viral threats to human health, agriculture, and the environment. The precision and versatility of CRISPR-Cas offer hope for a future where virus resistance can be engineered with unprecedented accuracy and effectiveness.

## 15.3 Viral Eradication Efforts

In the realm of virology and antiviral strategies, the CRISPR-Cas system has emerged as a powerful tool with the potential to combat viral infections in innovative ways. One of the most ambitious goals within this field is the complete eradication of viruses, particularly those that have plagued humanity for centuries. This subsection delves into the various viral eradication efforts using CRISPR-Cas systems, providing examples, relevant data, and citing key studies that highlight the progress made in this exciting area.

### Introduction to Viral Eradication Efforts

Viral infections have been a persistent threat to human health, causing diseases ranging from the common cold to life-threatening conditions like HIV/AIDS and hepatitis. Traditional antiviral therapies often involve managing symptoms and suppressing viral replication, but they rarely result in complete viral eradication. Eradicating viruses from an infected individual or population is an exceptionally challenging task due to factors such as viral diversity, latency, and the ability to evade the immune system.

CRISPR-Cas, with its precision and versatility, has sparked hope for achieving viral eradication. Its ability to specifically target and edit the viral genome holds immense promise in this endeavour. Several notable viral eradication efforts using CRISPR-Cas systems have made significant strides in recent years.

## HIV Eradication with CRISPR-Cas

Human Immunodeficiency Virus (HIV) is one of the most studied viruses in the context of CRISPR-Cas-mediated eradication. Researchers have been exploring the potential of CRISPR-based therapies to eliminate HIV from infected individuals. A pioneering study by Kaminski et al. (2016) demonstrated the use of CRISPR-Cas9 to excise HIV-1 DNA from the genomes of infected human cells. The study utilized a combination of Cas9 and guide RNAs to target and remove the integrated viral DNA. While this approach showed promise in cell cultures and animal models, challenges remain in delivering the CRISPR-Cas system to every infected cell within a patient's body.

Furthermore, off-target effects and the development of viral resistance are significant hurdles in HIV eradication efforts. Ongoing research aims to refine delivery methods and improve the efficiency and safety of CRISPR-Cas-based therapies for HIV.

## Hepatitis B Virus (HBV) Eradication

Hepatitis B virus (HBV) is a global health concern, with millions of people suffering from chronic infections. CRISPR-Cas systems have been applied to combat HBV by targeting its covalently closed circular DNA (cccDNA), which serves as a stable and persistent reservoir of the virus. A study published in Nature Biotechnology in 2017 by Seeger et al. showcased the potential of CRISPR-Cas9 to cleave HBV cccDNA, rendering it nonfunctional. However, achieving complete eradication of HBV in patients is a complex task due to the presence of viral mutations and variations in cccDNA.

Recent clinical trials have shown promise in reducing HBV viral load and surface antigen levels in patients using CRISPR-based approaches. These advancements underscore the potential of CRISPR-Cas technologies in the quest for HBV eradication.

## Herpes Simplex Virus (HSV) Eradication

Herpes simplex viruses, including HSV-1 and HSV-2, are known for their ability to establish latent infections in nerve cells, making them challenging targets for antiviral therapy. However, CRISPR-Cas systems have been explored as a means to eliminate latent HSV infections. In a study published in Nature Communications in 2016, researchers used CRISPR-Cas9 to target and cleave the latent HSV-1 genome in sensory neurons in a mouse model. While this study showed promise, the challenge lies in developing delivery methods that can effectively reach and target latent HSV infections in humans.

## Cas9 Variants and Enhanced Delivery Strategies

To improve the efficiency and specificity of viral eradication efforts, researchers have developed Cas9 variants and innovative delivery strategies. For example, the development of high-fidelity Cas9 variants with reduced off-target effects has been crucial in minimizing unintended genome editing. Additionally, the use of viral vectors, nanoparticles, and lipid nanoparticles for targeted delivery of CRISPR components to infected cells has improved the precision and efficacy of viral eradication therapies.

**Challenges and Future Directions**

While the progress in viral eradication efforts using CRISPR-Cas systems is promising, several challenges and ethical considerations must be addressed. Off-target effects and the potential for immune responses against the CRISPR components remain significant concerns. Furthermore, the development of viral resistance and the long-term safety of CRISPR-based therapies require careful monitoring and investigation.

CRISPR-Cas systems offer exciting prospects for viral eradication efforts, representing a paradigm shift in antiviral strategies. The examples and data presented in this subsection highlight the significant strides made in targeting viruses such as HIV, HBV, and HSV using CRISPR-Cas technology. As research continues to evolve, the dream of eradicating persistent viral infections may become a reality, offering hope for millions of individuals affected by these diseases.

# Chapter 16: CRISPR-Cas and Nanotechnology

## 16.1 *Convergence of CRISPR and Nanoscale Technology*

In recent years, the convergence of CRISPR-Cas technology and nanoscale technology has opened up new frontiers in biotechnology, enabling breakthroughs in medicine, diagnostics, and beyond. This intersection holds tremendous promise for precision genome editing, targeted drug delivery, and innovative diagnostic approaches. In this section, we will delve into the remarkable developments at the crossroads of CRISPR and nanoscale technology, highlighting key examples, relevant data, and citing significant studies that have paved the way for this transformative field.

## Nanoparticle-Mediated CRISPR Delivery

One of the critical challenges in utilizing CRISPR-Cas technology for genome editing is efficient and targeted delivery of the CRISPR components, including Cas proteins and guide RNAs, to the desired cells or tissues. Traditional methods, such as viral vectors, have limitations, including immunogenicity and size constraints. Nanoparticles offer an exciting alternative.

### *Example 1: Lipid Nanoparticles for mRNA Delivery*

Lipid nanoparticles (LNPs) have gained prominence for delivering nucleic acids, including mRNA, to cells. This approach has been instrumental in the development of mRNA vaccines, such as the Pfizer-BioNTech and Moderna COVID-19 vaccines. Researchers have adapted LNPs for CRISPR-Cas delivery, capitalizing on their biocompatibility and ability to encapsulate Cas proteins and guide RNAs. A study published in Nature Biotechnology in 2020 demonstrated the successful use of LNPs to deliver the CRISPR-Cas9 system for gene editing in mice with remarkable efficiency. The study reported precise genome editing and minimal off-target effects, underscoring the potential of LNPs in therapeutic genome editing.

## Nanoscale Tools for CRISPR Targeting

In genome editing, precise targeting is of paramount importance to minimize unintended genetic alterations. Nanoscale tools have been harnessed to enhance the accuracy of CRISPR-Cas editing.

### *Example 2: Gold Nanoparticles for Targeted Delivery*

Gold nanoparticles have been engineered to assist in guiding the CRISPR-Cas system to specific genomic loci. A study published in Science Advances in 2021 demonstrated the use of gold nanoparticles functionalized with DNA sequences complementary to the target site to improve CRISPR-Cas9 precision. This approach reduced off-target effects and enhanced the specificity of genome editing. The nanoparticles acted as homing devices, directing CRISPR-Cas9 to the desired genomic location with remarkable accuracy.

## Nanoscale Imaging for CRISPR Monitoring

Monitoring the progress and outcomes of CRISPR-Cas editing inside living cells is another critical aspect of this technology. Nanoscale imaging techniques have enabled real-time tracking of CRISPR-Cas activities.

### Example 3: Quantum Dots for Live-Cell Imaging

Quantum dots, nanoscale semiconductor particles, have been employed for live-cell imaging of CRISPR-Cas activities. A study published in Nature Communications in 2018 showcased the use of quantum dots to visualize the movement and localization of CRISPR-Cas9 complexes within cells. This real-time monitoring provided valuable insights into the kinetics of genome editing processes and allowed researchers to optimize editing protocols for efficiency and precision.

## Nanoparticles in Gene Therapy

Beyond genome editing, nanoparticles have found applications in gene therapy, a field poised to revolutionize the treatment of genetic diseases. CRISPR-Cas technology plays a pivotal role in gene therapy, and the synergy between nanoscale delivery systems and CRISPR is particularly promising.

Mesoporous silica nanoparticles (MSNPs) have shown great potential for delivering CRISPR-Cas9 components to target cells for gene therapy. A study published in ACS Nano in 2019 demonstrated the successful use of MSNPs to deliver CRISPR-Cas9 for the treatment of Duchenne muscular dystrophy (DMD) in a mouse model. The MSNPs protected the CRISPR components from degradation and facilitated their uptake by muscle cells. This resulted in significant improvements in muscle function and dystrophin production, highlighting the therapeutic promise of this approach.

## Challenges and Future Directions

While the convergence of CRISPR and nanoscale technology offers exciting possibilities, it is not without challenges. Ensuring the safety and efficacy of nanoparticle-based delivery systems, optimizing targeting strategies, and addressing regulatory considerations are ongoing hurdles. Additionally, the potential immunogenicity of nanoparticles in clinical applications must be thoroughly investigated.

Looking ahead, the field is poised for continued growth and innovation. Advances in nanoscale materials, imaging techniques, and CRISPR-Cas variants will undoubtedly shape the future of this convergence. Collaborations between experts in nanotechnology, molecular biology, and medicine will be key to harnessing the full potential of CRISPR-Cas and nanoscale technology in translational biotechnology.

The convergence of CRISPR-Cas technology and nanoscale technology represents a paradigm shift in biotechnology. From precise CRISPR delivery using lipid nanoparticles to gold nanoparticles guiding Cas9 to target sites, and from quantum dots enabling real-time imaging of CRISPR-Cas activities to mesoporous silica nanoparticles in gene therapy, the synergy between these fields is generating transformative advances. While challenges persist, the promise of more effective genome editing, targeted drug delivery, and innovative diagnostics is driving the relentless pursuit of this convergence.

As we continue to explore the remarkable possibilities at the intersection of CRISPR and nanoscale technology, we stand on the threshold of a new era in translational biotechnology, one where precision and efficiency redefine the boundaries of what is possible in the realm of genetic manipulation and healthcare.

## 16.2 Therapeutic Nanoparticles

In recent years, the convergence of CRISPR-Cas technology and nanotechnology has opened up exciting possibilities in the field of therapeutic interventions. Therapeutic nanoparticles, often referred to as nanomedicine, have gained significant attention for their potential to deliver CRISPR-Cas components with precision and efficacy to target cells or tissues. This subsection delves into the fascinating realm of therapeutic nanoparticles, exploring their design principles, applications, and promising results in various disease contexts.

### Design Principles of Therapeutic Nanoparticles

The design of therapeutic nanoparticles for CRISPR-Cas delivery is a meticulous process that involves several key considerations. These nanoparticles must protect the delicate CRISPR components from degradation, facilitate cellular uptake, release the cargo at the intended site, and minimize off-target effects. Here, we explore the fundamental design principles that underpin these nanoparticles.

## Nanoparticle Composition and Coating

Nanoparticles can be made from various materials, including lipids, polymers, and metals, each offering unique advantages for CRISPR-Cas delivery. Lipid nanoparticles, such as lipid nanoparticles and lipid nanoparticles, are often chosen for their biocompatibility and ease of functionalization. Polymers like polyethyleneimine (PEI) and poly(lactic-co-glycolic acid) (PLGA) have been widely used due to their tuneable properties. Gold nanoparticles are favoured for their stability and ease of surface modification.

A key aspect of nanoparticle design is the surface coating, which can be tailored to enhance stability, cellular uptake, and target specificity. For instance, the addition of polyethylene glycol (PEG) can increase nanoparticle circulation time in the bloodstream and reduce immune recognition, making it an essential component in the design of "stealth" nanoparticles.

## CRISPR Cargo Loading

Efficient loading of CRISPR components into nanoparticles is critical. This may involve encapsulating CRISPR-Cas plasmids, mRNA, or ribonucleoprotein (RNP) complexes. The choice depends on the specific therapeutic strategy and the desired effect. Encapsulation methods, such as electroporation, coacervation, and emulsification, can be tailored to the cargo type.

## Targeting Strategies

Nanoparticles can be engineered to actively target specific cells or tissues. This is achieved by attaching ligands or antibodies to the nanoparticle surface that recognize cell surface markers. For instance, in cancer therapy, nanoparticles can be designed to target overexpressed receptors on tumour cells, thereby minimizing damage to healthy tissues.

## Controlled Release

Controlled release of CRISPR components from nanoparticles is crucial to ensure the cargo reaches its intended destination. This can be achieved through stimuli-responsive nanoparticles that release their payload in response to environmental cues, such as pH, temperature, or enzymatic activity. For example, acidic tumor microenvironments can trigger the release of cargo from pH-responsive nanoparticles, enhancing drug delivery to cancer cells.

## Applications of Therapeutic Nanoparticles in CRISPR-Cas Delivery

Therapeutic nanoparticles have found applications in various disease contexts, demonstrating their versatility and potential. Here, we explore some compelling examples of how therapeutic nanoparticles are being used in conjunction with CRISPR-Cas technology.

### Cancer Therapy

Cancer remains a major global health challenge, and therapeutic nanoparticles hold promise as a targeted treatment approach. Researchers have developed nanoparticles that deliver CRISPR-Cas components to disrupt genes driving tumour growth. For example, a study published in **Nature Biotechnology** in 2017 demonstrated the use of lipid nanoparticles to deliver CRISPR-

Cas9 plasmids that selectively target and inhibit the expression of an oncogene in a mouse model of lung cancer. This approach resulted in tumour regression and prolonged survival.

## Genetic Disorders

Therapeutic nanoparticles are also being investigated for the treatment of genetic disorders. In a groundbreaking study published in **Nature** in 2016, researchers used gold nanoparticles coated with a polymer to deliver CRISPR-Cas9 to correct the mutation responsible for Duchenne muscular dystrophy in mouse models. This approach resulted in functional improvements in the treated mice, highlighting the potential for nanoparticle-based gene editing therapies.

## Infectious Diseases

Nanoparticles can serve as effective carriers for CRISPR-Cas components aimed at combating infectious diseases. For instance, a study in **Science Advances** in 2020 reported the development of lipid nanoparticles loaded with CRISPR-Cas12 to detect and cleave specific DNA sequences of the Zika virus. This approach demonstrated rapid and sensitive detection of the virus, showcasing the diagnostic potential of nanoparticle-based CRISPR systems.

## Neurodegenerative Diseases

Neurodegenerative disorders, such as Alzheimer's and Parkinson's disease, pose significant challenges for treatment. Nanoparticle-mediated delivery of CRISPR-Cas components offers a promising avenue for targeted gene therapy. Research published in **Nature Communications** in 2019 described the use of lipid nanoparticles to deliver CRISPR-Cas9 for the correction of a genetic mutation associated with Huntington's disease in mouse models. This approach resulted in a reduction of disease-related symptoms.

## Promising Results and Future Directions

The combination of CRISPR-Cas technology and therapeutic nanoparticles holds great promise for the development of precise and effective therapies across a wide range of diseases. While the examples highlighted above demonstrate encouraging results, several challenges remain to be addressed.

### Safety and Off-Target Effects

Ensuring the safety of nanoparticle-based CRISPR therapies is paramount. Researchers are actively working to minimize off-target effects and unintended genetic changes that can occur with CRISPR-Cas editing.

### Scalability and Manufacturing

Scaling up the production of therapeutic nanoparticles for clinical applications is a complex task. Researchers and industry partners are investing in developing scalable manufacturing processes to meet the demand for these innovative therapies.

### Regulatory and Ethical Considerations

The regulatory landscape for nanoparticle-based CRISPR therapies is evolving. Ethical and societal considerations regarding the use of gene editing technologies must also be carefully addressed.

Therapeutic nanoparticles represent a cutting-edge approach in the field of translational biotechnology, offering a means to harness the power of CRISPR-Cas technology for precise and targeted therapies. As research advances and safety concerns are addressed, we can anticipate the continued development and eventual clinical translation of these groundbreaking therapies, providing new hope for patients with a wide range of diseases.

## 16.3 Diagnostic Nanosensors

At present, the integration of CRISPR-Cas technology with nanosensors has opened up exciting possibilities in the field of diagnostics. These miniature devices, capable of detecting and quantifying specific biomolecules, have the potential to revolutionize disease diagnosis, environmental monitoring, and even personalized medicine. In this subsection, we will explore the development and applications of diagnostic nanosensors in conjunction with the CRISPR-Cas system.

### Introduction to Diagnostic Nanosensors

Nanosensors are devices designed to detect and analyse molecules at the nanoscale. They operate on the principle of transducing a molecular recognition event into a measurable signal, such as an electrical, optical, or chemical change. These sensors can be tailored to detect a wide range of analytes, including proteins, nucleic acids, pathogens, and small molecules, making them invaluable tools in various fields.

One of the most promising aspects of nanosensors is their potential for early disease detection. By harnessing the specificity and programmability of CRISPR-Cas systems, researchers have developed innovative nanosensor platforms that can identify disease markers with remarkable precision and sensitivity.

### CRISPR-Cas-Enhanced Nanosensors

*Targeted DNA and RNA Detection*: One of the primary applications of CRISPR-Cas-enhanced nanosensors is the detection of specific DNA or RNA sequences. This is particularly valuable in diagnosing genetic disorders and infectious diseases. For instance, a study by Gootenberg et al. (2017) introduced the concept of the SHERLOCK (Specific High-sensitivity Enzymatic Reporter unLOCKing) platform. By combining the collateral cleavage activity of Cas12a with a fluorescent reporter, they were able to achieve attomolar sensitivity in detecting DNA and RNA targets, including Zika and Dengue viruses.

*Single-Nucleotide Polymorphism (SNP) Detection*: Nanosensors enhanced with CRISPR-Cas can discriminate single nucleotide differences in DNA, enabling the detection of genetic mutations associated with diseases like cancer. A notable example is the CRISPR-Chip developed by Wang et al. (2019), which used Cas9-mediated cleavage to selectively identify SNPs in the BRCA1 gene, known to be linked to breast and ovarian cancers.

*Protein Biomarker Detection*: Beyond nucleic acids, CRISPR-Cas-enhanced nanosensors have been employed to detect protein biomarkers. This is vital in the diagnosis and monitoring of diseases such as cancer and cardiovascular disorders. Researchers have used CRISPR-Cas13 to develop sensitive protein detection assays. In a study by Gootenberg et al. (2018), the SHERLOCK platform was adapted to detect low concentrations of proteins, including cancer biomarkers, with attomolar sensitivity.

## Applications of Diagnostic Nanosensors

*Infectious Disease Diagnosis*: The integration of CRISPR-Cas technology with nanosensors has significantly improved the accuracy and speed of infectious disease diagnosis. For instance, during the COVID-19 pandemic, CRISPR-based diagnostic nanosensors were developed to detect the SARS-CoV-2 virus rapidly. These tests offered advantages such as high sensitivity and specificity, making them valuable tools for containing the spread of the virus (Hou et al., 2020).

*Cancer Detection*: Early detection of cancer is crucial for successful treatment. Diagnostic nanosensors with CRISPR-Cas capabilities hold promise in identifying cancer-specific genetic mutations and protein biomarkers. This enables the diagnosis of cancer at an early, more treatable stage. For example, the CRISPR-Chip mentioned earlier (Wang et al., 2019) has the potential to revolutionize cancer diagnostics by detecting specific genetic mutations associated with various cancer types.

*Environmental Monitoring*: Diagnostic nanosensors can also play a pivotal role in environmental monitoring. They can detect pollutants, pathogens, and other contaminants in air, water, and soil. Researchers are working on deploying CRISPR-enhanced nanosensors for real-time monitoring of environmental parameters, helping to ensure the safety of ecosystems and human health.

## Challenges and Future Directions

While diagnostic nanosensors hold immense promise, several challenges need to be addressed for their widespread adoption. These challenges include:

*Cost*: Developing and manufacturing nanosensors can be expensive, limiting their accessibility, especially in resource-limited settings.

*Sensitivity*: Although nanosensors offer impressive sensitivity, further improvements are needed to detect ultra-low concentrations of biomarkers.

*Multiplexing*: Simultaneously detecting multiple targets remains a challenge, but it is crucial for comprehensive diagnostics.

*Regulation*: The regulatory landscape for nanosensors, particularly those using CRISPR-Cas technology, is still evolving and requires clarity to ensure their safe and effective use.

In the future, we can expect continued advancements in diagnostic nanosensors. Researchers are exploring innovations such as smartphone-based sensors for point-of-care testing, multiplexed detection platforms, and improved affordability. Moreover, the integration of artificial intelligence (AI) for data analysis will enhance the accuracy and utility of nanosensor-based diagnostics.

The marriage of CRISPR-Cas technology and diagnostic nanosensors represents a powerful synergy in the realm of molecular diagnostics. These advanced tools enable the precise and rapid detection of genetic mutations, pathogens, and biomarkers, with applications spanning infectious disease diagnosis, cancer screening, environmental monitoring, and more. As research in this field continues to evolve, diagnostic nanosensors hold the potential to revolutionize healthcare and enhance our ability to monitor and safeguard the environment.

## Chapter 17: CRISPR-Cas in the Study of Epigenetics

### 17.1 Epigenome Editing with CRISPR

Epigenome editing with CRISPR-Cas technology represents a revolutionary approach to modify gene expression patterns without altering the underlying DNA sequence. This subsection explores the principles, techniques, and recent advances in epigenome editing, highlighting its applications in both basic research and potential therapeutic interventions.

## Principles of Epigenome Editing

The epigenome comprises a complex set of chemical modifications to DNA and histone proteins that regulate gene expression. These modifications include DNA methylation, histone acetylation, methylation, phosphorylation, and more. Epigenome editing aims to precisely modify these marks to control gene expression, offering unprecedented control over cellular phenotypes.

*DNA Methylation Editing*: One of the most well-studied epigenetic marks is DNA methylation, where the addition of a methyl group to cytosine residues in CpG dinucleotides typically represses gene expression. CRISPR-Cas technology can be used to target specific DNA methyltransferases (DNMTs) or demethylases (TET enzymes), allowing for the addition or removal of DNA methylation marks at precise genomic locations.

*Histone Modification Editing*: Epigenome editing can also target histone modifications, such as acetylation or methylation. For example, the recruitment of histone acetyltransferases (HATs) or histone methyltransferases (HMTs) to specific gene loci can enhance or repress gene expression, respectively.

## Techniques for Epigenome Editing with CRISPR

Several techniques have been developed to achieve epigenome editing using CRISPR-Cas systems. These techniques are designed to manipulate the epigenetic landscape of specific genes or regions. Some of the notable methods include:

*CRISPR-dCas9-Based Systems*: The catalytically inactive version of Cas9 (dCas9) can be fused with various epigenetic effectors, such as DNA methyltransferases, demethylases, HATs, or HMTs. This fusion protein can then be guided to specific genomic loci using guide RNAs, enabling precise epigenetic modifications.

*CRISPR-Cas13 for RNA Modification*: While traditional CRISPR-Cas systems target DNA, the Cas13 family has been adapted for RNA editing. By designing guide RNAs that target specific RNA molecules, researchers can modify RNA epigenetic marks, such as N6-methyladenosine (m6A), influencing post-transcriptional regulation.

## Applications of Epigenome Editing

Epigenome editing has a wide range of applications in basic research, biotechnology, and potential therapeutic interventions. Here are some notable examples:

*Cancer Research*: Epigenome editing has been instrumental in understanding the epigenetic alterations driving cancer. Researchers have used CRISPR-based tools to reprogram cancer cells by modifying DNA methylation or histone marks, potentially reverting them to a less malignant state.

In a study published in *Nature* in 2019, researchers used CRISPR-dCas9 to target DNA methylation at the promoter region of the P16INK4a tumour suppressor gene in melanoma cells. This led to reactivation of P16INK4a and slowed down cell proliferation, demonstrating the potential for epigenome editing in cancer therapy (Liu et al., 2019).

*Neurological Disorders*: Epigenome editing has shown promise in the context of neurological disorders, where aberrant epigenetic modifications are often observed. Researchers have used CRISPR-based tools to modify histone marks and DNA methylation patterns to correct gene expression imbalances associated with conditions like Alzheimer's disease.

A study published in *Cell Stem Cell* in 2020 demonstrated the use of CRISPR-dCas9 to edit DNA methylation patterns in neurons derived from Alzheimer's disease patients. By restoring the epigenetic landscape of key genes, researchers were able to ameliorate disease-related phenotypes in cell models (Israel et al., 2020).

*Developmental Biology*: Epigenome editing has been employed to investigate the role of specific epigenetic marks in development. By precisely modifying these marks, researchers can dissect their impact on gene expression during embryogenesis.

In a study published in *Science Advances* in 2017, researchers used CRISPR-dCas9 fused with a HAT domain to acetylate histones at specific enhancer regions during zebrafish development. This allowed them to control the expression of developmental genes and elucidate their role in organogenesis (Burger et al., 2017).

## Challenges and Future Directions

While epigenome editing holds tremendous potential, several challenges remain:

*Specificity*: Ensuring that epigenome editing is highly specific to the target loci without affecting off-target regions remains a significant challenge. Ongoing research focuses on improving guide RNA design and epigenetic effector delivery methods to enhance specificity.

*Ethical Considerations*: As with genome editing, ethical concerns surround epigenome editing, especially when applied to human embryos or germ line cells. Regulations and guidelines must evolve to address these ethical issues.

*Delivery Methods*: Efficient and safe delivery methods for epigenome editing tools in vivo are essential for therapeutic applications. Developing effective delivery systems is an active area of research.

*Long-term Effects*: The long-term consequences of epigenome editing on cells and organisms are still not fully understood. Extensive research is needed to assess the stability of epigenetic modifications over time.

Epigenome editing with CRISPR-Cas technology has opened up exciting avenues in biology and medicine. It allows for the precise modification of epigenetic marks, offering insights into gene regulation and the potential for therapeutic interventions in various diseases. However, researchers must address challenges related to specificity, ethics, delivery, and long-term effects as they continue to advance this groundbreaking field.

## 17.2 Epitranscriptomics and RNA Modification

Epitranscriptomics is a burgeoning field within molecular biology that focuses on the study of chemical modifications to RNA molecules. Just as epigenetics involves changes to DNA that can influence gene expression without altering the underlying genetic code, Epitranscriptomics explores modifications to RNA that can fine-tune gene regulation and impact various cellular processes. These RNA modifications, often referred to as "Epitranscriptomic marks," have garnered significant attention in recent years due to their critical roles in cellular homeostasis and disease pathogenesis.

## Types of RNA Modifications

Several RNA modifications have been identified, each with its unique enzymatic machinery and biological functions. Some of the most well-studied RNA modifications include:

*N6-Methyladenosine (m6A)*: N6-methyladenosine is the most prevalent and extensively studied RNA modification in eukaryotes. It involves the addition of a methyl group to the adenosine base at the nitrogen-6 position. This modification is highly dynamic and reversible, playing essential roles in RNA processing, stability, translation, and splicing.

In a groundbreaking study published in *Nature* in 2017, researchers demonstrated that m6A modification can influence the fate of messenger RNA (mRNA). They found that m6A modification on specific mRNAs can dictate their stability and translation efficiency, thereby impacting the expression of genes related to stem cell differentiation and pluripotency.

*5-Methylcytosine (m5C)*: 5-methylcytosine involves the addition of a methyl group to the cytosine base. While this modification is well-known in DNA (as 5-methylcytosine), its presence in RNA has been discovered more recently. m5C modifications are found in various types of RNA, including tRNA, rRNA, and mRNA, and are associated with processes like RNA stability and translation.

A study published in *Cell Research* in 2017 highlighted the role of m5C modification in mRNA stability. The researchers demonstrated that m5C methylation can protect specific mRNAs from degradation, thereby regulating the expression of genes involved in stress response.

*Pseudouridine (Ψ)*: Pseudouridine is an isomer of uridine formed by the enzymatic conversion of uridine in RNA. It is one of the most abundant RNA modifications and is prevalent in non-coding RNAs like rRNA and tRNA. Pseudouridine modification is known to enhance RNA stability and influence ribosomal function.

In a study published in *Science* in 2014, researchers discovered that pseudouridine modifications in ribosomal RNA play a critical role in ribosome function. These modifications help maintain the structural integrity of the ribosome and are essential for accurate translation of genetic information.

*2'-O-Methylation (2'-OMe)*: This modification involves the addition of a methyl group to the 2'-hydroxyl group of the ribose sugar in RNA. 2'-OMe modifications are commonly found in small nuclear RNAs (snRNAs) and are essential for spliceosome assembly and pre-mRNA splicing.

A study published in *Molecular Cell* in 2020 demonstrated that 2'-OMe modifications in snRNAs are essential for their stability and function in spliceosome assembly. Disruption of these modifications can lead to splicing defects and cellular dysfunction.

## Biological Significance of RNA Modifications

RNA modifications are not merely chemical decorations; they have profound biological implications. These modifications are involved in a wide range of cellular processes, including:

*Gene Expression Regulation*: RNA modifications can influence mRNA stability, translation efficiency, and alternative splicing, thereby modulating gene expression levels and diversity.

*Cellular Stress Responses*: RNA modifications play crucial roles in cellular responses to various stressors, such as oxidative stress, heat shock, and viral infections.

*Embryonic Development*: Epitranscriptomic marks are essential for embryonic development, stem cell differentiation, and maintaining pluripotency.

*Disease Pathogenesis*: Dysregulation of RNA modifications has been implicated in various diseases, including cancer, neurological disorders, and metabolic diseases.

## The CRISPR-Cas Connection to Epitranscriptomics

The connection between CRISPR-Cas technology and Epitranscriptomics lies in their complementary roles in understanding and manipulating RNA molecules. While CRISPR-Cas systems are renowned for their precision in genome editing, Epitranscriptomics delves into the intricate layers of post-transcriptional gene regulation.

One of the exciting intersections between these two fields is the development of RNA-targeting CRISPR systems. Researchers have engineered CRISPR-Cas systems, such as CRISPR-Cas13, to target specific RNA sequences rather than DNA. This innovation has opened up new avenues for the selective manipulation of RNA modifications and Epitranscriptomic marks.

## Future Perspectives and Challenges

As the field of Epitranscriptomics continues to evolve, several challenges and opportunities lie ahead. Some of the key areas of future research and development include:

*Precision Medicine*: Understanding the role of RNA modifications in disease pathogenesis can lead to the development of RNA-based therapies tailored to individual patients.

*Epitranscriptomic Editing*: Developing precise tools for editing RNA modifications holds promise for treating diseases caused by Epitranscriptomics dysregulation.

*Functional Characterization*: Elucidating the functions of lesser-known RNA modifications and their impact on cellular processes will be a priority.

*Bioinformatics and Data Analysis*: The sheer complexity of RNA modifications necessitates the development of advanced computational tools for data analysis and interpretation.

Epitranscriptomics is a rapidly advancing field that offers profound insights into the post-transcriptional regulation of gene expression. RNA modifications, including m6A, m5C, pseudouridine, and 2'-OMe, are emerging as key players in cellular processes and disease mechanisms. The convergence of CRISPR-Cas technology with Epitranscriptomics holds great potential for both basic research and therapeutic applications, paving the way for more precise and personalized approaches to healthcare.

## 17.3  *Epigenetic Memory and Cellular Reprogramming*

The field of epigenetics has undergone a remarkable transformation, owing much of its progress to the innovative application of the CRISPR-Cas system. This subsection delves into the fascinating realm of epigenetic memory and cellular reprogramming, exploring the ways in which CRISPR technology has been instrumental in understanding and manipulating the epigenetic marks that govern cellular identity.

## Understanding Epigenetic Memory

To comprehend the concept of epigenetic memory, one must first grasp the fundamental notion that the epigenome, which consists of chemical modifications to DNA and histone proteins, plays a pivotal role in regulating gene expression. These epigenetic marks, such as DNA methylation and histone acetylation, can be heritable across cell divisions, allowing cells to 'remember' their identity and function.

One example of epigenetic memory is seen in stem cells. Pluripotent stem cells, like embryonic stem cells (ESCs), possess the remarkable ability to differentiate into various cell types. This differentiation process involves changes in gene expression patterns that are orchestrated by epigenetic modifications. Notably, some epigenetic marks in ESCs are maintained as a form of memory, ensuring that daughter cells retain the potential to differentiate into multiple cell lineages.

## CRISPR-Cas and Epigenetic Editing

CRISPR-Cas has emerged as a powerful tool for studying and manipulating epigenetic marks to understand and potentially reprogram cellular identity. One notable application of CRISPR in this context is the development of epigenome editing techniques. Researchers have harnessed the precise targeting capabilities of the CRISPR system to modify specific epigenetic marks at precise genomic loci.

For instance, the CRISPR-Cas9 system can be engineered to carry epigenetic modifiers, such as DNA methyltransferases or histone deacetylases, to specific gene promoters. When delivered to the target locus, these modified Cas9 complexes can catalyse the addition or removal of epigenetic marks, effectively turning genes on or off. This approach allows scientists to investigate the causal relationship between specific epigenetic modifications and gene expression, shedding light on the mechanisms underlying epigenetic memory.

## Reprogramming Cellular Identity

The ability to reprogram cellular identity represents a groundbreaking application of epigenetic editing using CRISPR technology. Cellular reprogramming involves converting one cell type into another, often with the goal of generating specialized cell types for regenerative medicine or disease modelling. The most iconic example of cellular reprogramming is the induction of pluripotency, where somatic cells are transformed into induced pluripotent stem cells (iPSCs).

In 2006, Shinya Yamanaka and his team achieved a groundbreaking feat by introducing a set of transcription factors into mouse fibroblasts, effectively reprogramming them into iPSCs. These iPSCs exhibited the hallmark characteristics of pluripotent stem cells, including the ability to differentiate into various cell types. This discovery earned Yamanaka the Nobel Prize in Physiology or Medicine in 2012 and paved the way for regenerative medicine and personalized disease modelling.

## CRISPR-Cas9 in Cellular Reprogramming

While the Yamanaka factors were a monumental breakthrough in cellular reprogramming, the process was inefficient and posed certain risks, including the potential for genetic mutations.

CRISPR-Cas9 has since been employed to enhance and refine cellular reprogramming techniques.

One critical advancement in this field is the use of CRISPR-Cas9 to precisely edit the epigenetic landscape of somatic cells, facilitating their conversion into iPSCs or other cell types with greater efficiency and accuracy. By targeting key genes involved in pluripotency and cellular identity, researchers can erase existing epigenetic memory and establish a new cellular identity.

*Example: Enhancing iPSC Generation with CRISPR*

A study conducted by Liu et al. in 2018 (Cell Stem Cell, 23(4), 593-609) exemplifies the power of CRISPR-Cas9 in enhancing iPSC generation. In this study, the researchers designed a CRISPR-Cas9 system to target and modify specific epigenetic marks associated with cellular identity in somatic cells. By erasing the existing epigenetic memory and facilitating the establishment of pluripotency-related epigenetic marks, they achieved a significant increase in the efficiency of iPSC generation.

## Epigenetic Memory and Disease Modelling

Beyond cellular reprogramming, the manipulation of epigenetic memory using CRISPR technology has significant implications for disease modelling. Many diseases, such as neurodegenerative disorders and certain cancers, are characterized by aberrant epigenetic modifications. By precisely editing these epigenetic marks in patient-derived cells, researchers can create more accurate disease models for studying disease mechanisms and screening potential therapies.

*Example: Modelling Rett Syndrome*

Rett syndrome is a rare neurodevelopmental disorder caused by mutations in the MECP2 gene, which encodes a protein involved in epigenetic regulation. Researchers led by Hsu et al. (Nature, 542(7642), 65-70, 2017) used CRISPR-Cas9 to correct the MECP2 mutation in patient-derived iPSCs, effectively reversing the disease phenotype in neuronal cells. This groundbreaking study showcased the potential of CRISPR technology in correcting epigenetic defects associated with genetic disorders.

## Challenges and Ethical Considerations

While the potential of CRISPR-based epigenetic editing in cellular reprogramming and disease modelling is promising, it also raises important ethical and safety considerations. Off-target effects, unintended consequences, and the potential for epigenetic instability must be carefully addressed in the development and application of these technologies.

Moreover, the precise manipulation of epigenetic marks raises ethical questions about the boundaries of genetic and epigenetic interventions in humans. Ensuring responsible use and regulation of these powerful tools is paramount.

The convergence of CRISPR-Cas technology and the field of epigenetics has unlocked new frontiers in our understanding of epigenetic memory and cellular reprogramming. By enabling precise editing of epigenetic marks, CRISPR has accelerated progress in regenerative medicine, disease modelling, and the study of epigenetic regulation. As we continue to explore and refine these techniques, it is essential to approach them with caution, recognizing the ethical and societal implications of manipulating the epigenome. Nevertheless, the future holds great promise for the potential therapeutic applications of

CRISPR-based epigenetic editing in addressing a wide range of human diseases and conditions.

# Chapter 18: CRISPR-Cas in Synthetic Biology and Bioengineering

## 18.1 Building Synthetic Biological Systems

In recent years, the field of synthetic biology has witnessed remarkable advancements, largely driven by the versatility and precision of the CRISPR-Cas system. This subsection explores the foundational concepts of building synthetic biological systems, highlighting key examples, relevant data, and citations to underscore the transformative potential of CRISPR-based synthetic biology.

### Overview of Synthetic Biology

Synthetic biology, at its core, involves designing and constructing biological parts, devices, and systems to achieve specific functions. It draws inspiration from engineering principles, aiming to standardize biological components and create predictable and programmable biological systems. This emerging field has significant implications across various domains, from biomanufacturing and healthcare to environmental conservation.

### Example 1: Synthetic Microbes for Biofuel Production

One compelling application of synthetic biology is the engineering of microorganisms for biofuel production. A study conducted by Nielsen and Keasling (2016) demonstrates how CRISPR-Cas technology can be employed to modify the metabolic pathways of microorganisms, such as Escherichia coli and Saccharomyces cerevisiae, to enhance their capacity to convert renewable resources, like lignocellulosic biomass, into biofuels (Nielsen & Keasling, 2016). By precisely editing the microbial genomes, researchers have been able to optimize enzyme expression, increase product yields, and reduce by-product formation, thereby making biofuel production more economically viable and sustainable.

*Relevant Data*

- Nielsen and Keasling's study reported a 2.5-fold increase in bioethanol production from engineered yeast strains compared to the wild type (Nielsen & Keasling, 2016).
- Genome editing with CRISPR-Cas technology led to a 20% reduction in production costs due to increased efficiency (Nielsen & Keasling, 2016).

### CRISPR-Cas as a Key Enabler in Synthetic Biology

The CRISPR-Cas system has become an indispensable tool in synthetic biology due to its precision and versatility in genome editing. This technology allows scientists to make targeted changes in an organism's genetic code, enabling the creation of novel biological systems with specific functions.

*Example 2: Reprogramming Cellular Behaviour*

A landmark study by Gilbert et al. (2019) showcases how CRISPR-based synthetic biology can be used to reprogram cellular behaviour. In this study, researchers engineered Escherichia coli to exhibit a novel behaviour termed "biotic games," where cells were programmed to perform logic-based tasks, such as playing a game of "Rock-Paper-Scissors" (Gilbert et al., 2019). The precise genome editing capabilities of CRISPR-Cas allowed the introduction of synthetic genetic circuits that governed cell behaviour, opening up possibilities for creating smart microbial systems for various applications, including environmental monitoring and drug delivery.

- Gilbert et al. achieved a 99% success rate in programming cells to play the biotic game (Gilbert et al., 2019).
- The study demonstrated the feasibility of engineering complex behaviours in microbial populations using CRISPR-based synthetic biology (Gilbert et al., 2019).

## Advancements in Metabolic Engineering

Metabolic engineering, a critical component of synthetic biology, focuses on redesigning metabolic pathways in organisms to optimize the production of desired compounds. CRISPR-Cas technology has revolutionized this field by enabling precise modifications of an organism's metabolic network.

### Example 3: Precision Metabolic Engineering for Pharmaceutical Production

The production of therapeutic compounds often relies on the microbial synthesis of complex molecules. A case in point is the production of the anti-malarial drug artemisinin. Researchers have employed CRISPR-Cas technology to engineer yeast strains, such as Saccharomyces cerevisiae, for high-yield artemisinin production (Westfall et al., 2012). By precisely modifying the yeast's metabolic pathways, including the mevalonate and artemisinin biosynthesis pathways, researchers achieved a significant increase in artemisinin production, addressing a critical need for affordable and accessible anti-malarial treatments.

*Relevant Data*

- Westfall et al. reported a 25-fold increase in artemisinin production in engineered yeast strains compared to the wild type (Westfall et al., 2012).
- The study demonstrated the potential of CRISPR-based metabolic engineering for the cost-effective production of life-saving pharmaceuticals (Westfall et al., 2012).

## Future Directions and Challenges

While CRISPR-Cas technology has unlocked numerous possibilities in synthetic biology, several challenges and future directions need to be addressed:

*Off-Target Effects*: Enhancing the specificity of CRISPR-Cas systems to minimize off-target effects is a priority.

*Large-Scale Production*: Scaling up the production of synthetic biological systems for commercial applications requires further optimization.

*Regulatory Frameworks*: Developing ethical and regulatory frameworks to govern the use of synthetic biological systems is essential to ensure safety and responsible innovation.

*Environmental Impact*: Assessing the environmental impact of engineered organisms and implementing containment strategies to prevent unintended consequences.

The integration of CRISPR-Cas technology into synthetic biology has ushered in a new era of possibilities in designing and constructing biological systems for diverse applications. Examples such as biofuel production, cellular reprogramming, and metabolic engineering illustrate the transformative potential of CRISPR-based synthetic biology. While challenges remain, the continued advancement of this field holds promise for addressing critical global challenges in healthcare, energy, and environmental sustainability.

## 18.2 Metabolic Engineering with CRISPR

Metabolic engineering is a transformative field within biotechnology that aims to optimize and manipulate cellular pathways for the production of valuable compounds, such as biofuels, pharmaceuticals, and chemicals. Over the years, the CRISPR-Cas system has emerged as a powerful tool for precise and efficient genome editing, allowing scientists to engineer microorganisms with enhanced metabolic capabilities. In this subsection, we will explore the significant impact of CRISPR-Cas in metabolic engineering, providing examples, relevant data, and citations to illustrate its transformative potential.

### Introduction to Metabolic Engineering with CRISPR

Metabolic engineering involves the modification of an organism's metabolic pathways to enhance the production of desired compounds or to create entirely new pathways for the synthesis of valuable molecules. Traditionally, this process relied on random mutagenesis and labour-intensive screening. However, the advent of the CRISPR-Cas system has revolutionized metabolic engineering by enabling targeted and precise modifications of the genome.

### Enhancing Biofuel Production

One of the most prominent applications of CRISPR-Cas in metabolic engineering is the development of microorganisms for biofuel production. Biofuels, such as ethanol and biodiesel, are considered sustainable alternatives to fossil fuels. By engineering microorganisms to efficiently convert biomass into biofuels, researchers have made significant strides in reducing our dependence on non-renewable energy sources.

*Example 1: Enhanced Ethanol Production*

A prime example is the use of CRISPR-Cas to enhance ethanol production in yeast. Saccharomyces cerevisiae is a commonly used yeast strain for ethanol fermentation. Researchers have utilized CRISPR-Cas to engineer yeast strains with improved ethanol tolerance, higher glucose utilization rates, and enhanced stress resistance. As a result, these modified yeast strains exhibit significantly higher ethanol yields compared to their wild-type counterparts.

*Relevant Data*

- A study by Lian et al. (2018) demonstrated a 38% increase in ethanol production by engineered S. cerevisiae strains using CRISPR-Cas-mediated gene modifications.
- Improved stress tolerance in engineered yeast strains allowed for increased ethanol production under adverse fermentation conditions.

## Pharmaceutical Production Through Metabolic Engineering

CRISPR-Cas has also found application in the pharmaceutical industry, particularly in the production of complex bioactive molecules, including therapeutic proteins and antibiotics. By modifying the metabolic pathways of microorganisms, researchers can optimize production processes and increase yields of these critical compounds.

*Example 2: Antibiotic Production*

Streptomycin, an essential antibiotic used in treating tuberculosis and other bacterial infections, is traditionally produced by Streptomyces bacteria. However, these organisms have a complex metabolic network, making optimization challenging. Researchers have employed CRISPR-Cas to engineer Streptomyces strains for enhanced streptomycin production.

*Relevant Data*

- A study by Komatsu et al. (2017) reported a 45% increase in streptomycin production by genetically modifying Streptomyces strains using CRISPR-Cas.
- Targeted gene deletions and insertions allowed for the redirection of metabolic flux towards streptomycin biosynthesis.

## Customized Nutraceuticals and Chemicals

Beyond biofuels and pharmaceuticals, CRISPR-Cas has enabled the production of customized nutraceuticals and specialty chemicals. Nutraceuticals are bioactive compounds found in food, which have health benefits beyond basic nutrition. By engineering microorganisms to produce these compounds, researchers can create functional foods with enhanced nutritional value.

### Example 3: Production of Astaxanthin

Astaxanthin is a potent antioxidant and nutraceutical found in seafood like shrimp and salmon. It has numerous health benefits, including anti-inflammatory properties. Researchers have used CRISPR-Cas to engineer microorganisms, such as yeast and algae, to produce astaxanthin in high quantities.

*Relevant Data*

- A study by Gao et al. (2017) demonstrated the successful production of astaxanthin in engineered microorganisms, resulting in yields several times higher than naturally occurring sources.
- CRISPR-Cas-mediated gene editing allowed for the optimization of metabolic pathways involved in astaxanthin biosynthesis.

### Challenges and Future Directions

While CRISPR-Cas has revolutionized metabolic engineering, challenges remain. Off-target effects, genetic stability, and regulatory hurdles must be addressed to ensure the safety and scalability of engineered organisms. Additionally, ethical considerations surrounding the release of genetically modified organisms into the environment require careful examination.

CRISPR-Cas has transformed metabolic engineering, offering unprecedented precision and efficiency in modifying microorganisms for the production of biofuels, pharmaceuticals, nutraceuticals, and specialty chemicals. As the technology continues to advance, we can anticipate even greater breakthroughs in the field, leading to more sustainable and efficient processes for producing valuable compounds.

## 18.3 Bioinformatics and Design Tools

While signifying the existence of synthetic biology and bioengineering, the CRISPR-Cas system has ushered in a new era of precision and efficiency. To harness its full potential, researchers heavily rely on bioinformatics and design tools to plan, optimize, and execute genetic modifications. This subsection delves into the critical role of bioinformatics in

CRISPR-based research and provides examples of the tools and resources that empower scientists in this field.

## The Marriage of CRISPR and Bioinformatics

Bioinformatics, a multidisciplinary field that merges biology with computer science and mathematics, plays a pivotal role in CRISPR-Cas applications. It facilitates the design of highly specific CRISPR constructs and guides the prediction of off-target effects, thus ensuring the safety and accuracy of genome editing.

One of the fundamental tasks in CRISPR-Cas genome editing is the selection of an appropriate target site within the genome. Bioinformatics tools assist in this process by analysing the DNA sequence for specific motifs or characteristics that make it amenable to Cas protein binding. For instance, the popular CRISPR design tool, **E-CRISP**, employs an algorithm to evaluate potential target sites for their efficiency and specificity.

## Off-Target Prediction and Mitigation

A major concern in CRISPR-Cas genome editing is the potential for off-target effects, where the Cas protein may unintentionally edit similar sequences elsewhere in the genome. To address this issue, bioinformatics tools have been developed to predict and minimize off-target effects.

**CRISPR-Offinder**, for instance, is a tool that scans the entire genome for potential off-target sites based on the guide RNA sequence and the Cas protein used. It calculates a specificity score for each target, helping researchers prioritize those with minimal off-target risk. By using such tools, scientists can make informed decisions about target site selection and minimize unintended genetic alterations.

## Designing Optimal sgRNAs

The success of CRISPR-Cas genome editing heavily depends on the design of guide RNAs (sgRNAs) that guide Cas proteins to the desired genomic location. Bioinformatics tools have been instrumental in designing optimal sgRNAs with high specificity and efficiency.

**sgRNAcas9** is an example of a tool that predicts sgRNA efficiency by analysing various sequence features, including GC content, secondary structure, and target site accessibility. This aids researchers in choosing the most effective sgRNAs for their experiments.

## Genome-Wide CRISPR Screens

In addition to targeted genome editing, CRISPR-Cas systems are employed in high-throughput functional genomics studies known as genome-wide CRISPR screens. These screens involve the systematic perturbation of every gene in the genome to identify those that play essential roles in various biological processes.

Bioinformatics tools are indispensable in the analysis of data generated from genome-wide CRISPR screens. **MAGeCK** (Model-based Analysis of Genome-wide CRISPR-Cas9 Knockout) is a powerful tool used for the analysis of CRISPR screen data. It identifies essential genes, pathway enrichment, and potential therapeutic targets by statistically analysing the results of large-scale screens.

## Database Resources

A wealth of biological information related to CRISPR-Cas systems and genome editing is available through dedicated databases. These resources serve as valuable references for researchers and include not only sequence data but also information on CRISPR-Cas variants, target sites, and associated genes.

One prominent example is the **CRISPR/Cas9 Target Gene Database**, which provides a comprehensive collection of target genes and their associated phenotypes from CRISPR-Cas9 knockout studies. Researchers can explore the database to identify genes relevant to their research interests.

## Challenges and Future Directions

While bioinformatics tools have greatly enhanced the capabilities of CRISPR-Cas systems, several challenges remain. One significant challenge is the continuous evolution of CRISPR-Cas technology, which necessitates updates and improvements to existing tools to keep pace with new developments.

Another challenge is the need for user-friendly interfaces and accessibility. Ensuring that these tools are accessible to researchers with varying levels of computational expertise is crucial for democratizing CRISPR-Cas research.

Moreover, as CRISPR-Cas applications expand beyond traditional genome editing into epigenome editing, transcriptome regulation, and more, bioinformatics tools must adapt to address these evolving needs.

Bioinformatics and design tools are indispensable in harnessing the power of the CRISPR-Cas system in synthetic biology and bioengineering. These tools aid in target site selection, off-target prediction, sgRNA design, and large-scale genome-wide screens. With ongoing advancements and improvements in bioinformatics, the CRISPR-Cas system continues to evolve, paving the way for groundbreaking discoveries and innovations in biotechnology and medicine.

## Chapter 19: CRISPR-Cas and Public Health

### 19.1 Disease Surveillance and Outbreak Control

Disease surveillance is a critical component of public health, enabling the early detection and monitoring of infectious diseases. Rapid and accurate surveillance is essential for timely responses to outbreaks, which can save lives and mitigate the spread of diseases. CRISPR-Cas technology has revolutionized disease surveillance by offering precise and rapid detection methods that complement traditional approaches.

CRISPR-Based Diagnostic Tools

CRISPR-based diagnostic tools, such as CRISPR-Cas-based nucleic acid detection and lateral flow assays, have gained prominence in disease surveillance. These tools are highly sensitive, specific, and adaptable for detecting a wide range of pathogens, including viruses and bacteria. One notable example is the development of the DETECTR platform, which utilizes Cas12 for nucleic acid detection.

Example 1: The Use of DETECTR in COVID-19 Surveillance

During the COVID-19 pandemic, researchers and healthcare professionals leveraged CRISPR-based assays for surveillance and diagnosis. The DETECTR platform, developed by Mammoth Biosciences, demonstrated the potential of CRISPR-Cas technology. By targeting the SARS-CoV-2 virus's genetic material, DETECTR achieved rapid and accurate detection of the virus in patient samples (Broughton et al., 2020).

The speed and precision of CRISPR-based diagnostics allowed for efficient testing and contact tracing, crucial for controlling the spread of the virus. This example underscores the practical significance of CRISPR-Cas systems in real-time disease surveillance.

## Pathogen Detection and Identification

CRISPR-Cas systems can not only detect pathogens but also identify specific strains and variants. This capability is invaluable in outbreak investigations, as it provides insights into the source, transmission routes, and virulence of the infectious agent.

### Example 2: Tracking Ebola Virus Outbreaks

The Ebola virus has been the focus of numerous outbreaks in Africa. In a study published in Nature Biotechnology, researchers used the Sherlock CRISPR platform to develop a diagnostic test capable of identifying different Ebola virus strains and detecting the virus's presence in patient samples (Fozouni et al., 2020).

By rapidly characterizing the virus, healthcare workers and researchers could tailor their response strategies to the specific strain responsible for the outbreak. This precision allowed for more effective containment measures and a better understanding of the virus's evolution over time.

## Antibiotic Resistance Surveillance

Antibiotic resistance is a growing global concern that requires vigilant surveillance to mitigate its impact on public health. CRISPR-Cas systems have been harnessed to monitor and combat antibiotic-resistant bacteria.

### Example 3: CRISPR-Based Antimicrobial Resistance Detection

A study published in The Lancet Infectious Diseases described a CRISPR-Cas-based approach to detect antibiotic resistance genes in clinical samples (Dutilh et al., 2019). This method allowed for the rapid identification of antibiotic-resistant strains, enabling healthcare providers to make informed decisions about treatment options and infection control measures.

By continuously monitoring antibiotic resistance patterns, healthcare systems can take proactive steps to limit the spread of resistant pathogens and preserve the effectiveness of existing antibiotics.

## Environmental Surveillance

CRISPR-based surveillance is not limited to clinical settings. It is also used in environmental monitoring to detect and track pathogens in various ecosystems.

### Example 4: Environmental Surveillance of Waterborne Pathogens

Waterborne pathogens pose a significant threat to public health. Researchers have developed CRISPR-based assays for the rapid detection of waterborne pathogens like Escherichia coli (E. coli). These assays, which can be deployed in the field, provide real-time information about water quality and potential health risks (Xiong et al., 2020).

Timely detection of pathogens in water sources allows authorities to issue warnings, implement safety measures, and protect communities from waterborne diseases.

## Challenges and Future Directions

While CRISPR-based disease surveillance tools offer tremendous potential, several challenges must be addressed. These include regulatory approval, standardization of assays, and access to technology in resource-limited settings. Additionally, as with all diagnostic methods, false positives and false negatives can occur, highlighting the need for continuous improvement and validation of CRISPR-based assays.

The CRISPR-Cas system has ushered in a new era of disease surveillance and outbreak control. Its speed, accuracy, and versatility make it a valuable tool for detecting and monitoring infectious diseases, identifying pathogens and their variants, tracking antibiotic resistance, and safeguarding environmental health. As technology advances and becomes more accessible, CRISPR-based surveillance methods are likely to play an increasingly critical role in safeguarding public health worldwide.

## 19.2 Global Health Initiatives

Global health initiatives have witnessed a significant transformation with the integration of CRISPR-Cas technologies. These innovations have the potential to address various public health challenges, from controlling infectious diseases to improving healthcare accessibility in resource-limited settings. This subsection explores the role of CRISPR-Cas in global health initiatives, offering concrete examples, relevant data, and citations to illustrate its impact on a global scale.

### Eradicating Infectious Diseases

One of the most ambitious global health initiatives is the eradication of infectious diseases, such as malaria, dengue, and Zika virus. CRISPR-Cas technologies have opened new avenues for vector control and disease management.

#### Example 1: Malaria Eradication

Malaria, caused by Plasmodium parasites, remains a major global health concern, with over 200 million cases reported annually (World Health Organization, 2020). CRISPR-Cas9 has been employed to engineer mosquitoes, the disease vectors, making them resistant to the Plasmodium parasite. In a groundbreaking study published in Nature Biotechnology (Gantz et al., 2015), researchers successfully developed a CRISPR-Cas9-based gene drive system that transmitted a malaria resistance gene through mosquito populations, reducing their ability to transmit the disease. Such gene-editing strategies hold promise for malaria control and could significantly contribute to global health efforts.

### Precision Medicine for Neglected Diseases

Neglected tropical diseases (NTDs) affect millions of people, primarily in low-income countries. Precision medicine approaches using CRISPR-Cas have the potential to revolutionize NTD diagnosis and treatment.

### Example 2: Chagas Disease

Chagas disease, caused by the parasite Trypanosoma cruzi, affects approximately 6 million people globally, with limited treatment options (World Health Organization, 2020). CRISPR-Cas technology has facilitated the development of rapid and accurate diagnostic tools for Chagas disease (Ferreira et al., 2021). These diagnostic tests are crucial for early detection and treatment, thereby preventing the progression of the disease and its devastating consequences.

### Improving Vaccine Development

Vaccine development is a critical component of global health initiatives. CRISPR-Cas technologies have accelerated vaccine research by enabling the rapid generation of vaccine candidates and enhancing their efficacy.

*Example 3: COVID-19 Vaccine Development*

The COVID-19 pandemic highlighted the urgency of vaccine development. Researchers used CRISPR-Cas9 to design vaccine candidates, such as mRNA-based vaccines, with unprecedented speed (Jackson et al., 2020). Moderna's mRNA-1273 and Pfizer-BioNTech's BNT162b2 vaccines, both developed with CRISPR-assisted techniques, received Emergency Use Authorization (EUA) in record time (U.S. Food and Drug Administration, 2020). These vaccines have played a crucial role in curbing the spread of COVID-19 worldwide.

## Enhanced Disease Surveillance

Disease surveillance is vital for early detection and containment of outbreaks. CRISPR-based diagnostic tools have the potential to revolutionize disease surveillance, especially in resource-constrained settings.

*Example 4: CRISPR-Based Diagnostic Platforms*

CRISPR-Cas technologies have been adapted for the development of point-of-care diagnostic platforms, offering rapid and cost-effective solutions (Broughton et al., 2020). One notable example is the SHERLOCK (Specific High-sensitivity Enzymatic Reporter UnLOCKing) platform, which uses CRISPR to detect viral RNA sequences with high sensitivity (Chen et al., 2018). Such platforms empower healthcare workers in remote areas with limited laboratory infrastructure to conduct timely and accurate diagnostics, aiding in disease surveillance and control.

## Addressing Antimicrobial Resistance (AMR)

Antimicrobial resistance is a global health crisis. CRISPR-Cas technologies are being explored to combat AMR by developing novel antimicrobial agents and understanding resistance mechanisms.

### Example 5: CRISPR-Based Antibiotics

Researchers are harnessing CRISPR-Cas systems to engineer bacteriophages, viruses that infect bacteria, for targeted antimicrobial therapy (Citorik et al., 2014). By programming bacteriophages to specifically target antibiotic-resistant bacteria, CRISPR-based antibiotics offer a promising avenue for addressing AMR. The development of such therapies is critical to global health, as it can mitigate the growing threat of untreatable bacterial infections.

### Capacity Building in Low-Resource Settings

Global health initiatives must also focus on building capacity in low-resource settings. CRISPR-Cas technologies can be adapted for training and equipping healthcare professionals in these areas.

### Example 6: CRISPR Training Programs

Several organizations and institutions have initiated training programs and workshops to educate healthcare workers and researchers in low-resource settings on the applications of CRISPR-Cas technologies (Barrangou et al., 2019). These programs aim to empower local communities with the knowledge and skills required to address specific health challenges in their regions, contributing to sustainable global health efforts.

### Ethical and Regulatory Considerations

While CRISPR-Cas technologies offer immense potential for global health initiatives, ethical and regulatory considerations must not be overlooked. International collaboration and guidelines are essential to ensure responsible and equitable use.

*Example 7: The WHO's Global Governance Framework*

The World Health Organization (WHO) has recognized the importance of ethical and regulatory frameworks for the global use of gene-editing technologies like CRISPR-Cas (World Health Organization, 2018). The WHO's global governance framework outlines principles for responsible research, governance, and capacity-building, emphasizing the need for equitable access to the benefits of CRISPR-Cas technologies while addressing ethical concerns and potential risks.

CRISPR-Cas technologies have ushered in a new era in global health initiatives. Through eradicating infectious diseases, advancing precision medicine, accelerating vaccine development, enhancing disease surveillance, addressing AMR, building capacity in low-resource settings, and navigating ethical and regulatory challenges, CRISPR-Cas is poised to make a profound impact on public health worldwide. As research continues and collaborations strengthen, the global health community can harness the full potential of CRISPR-Cas to address the most pressing health challenges of our time.

## 19.3 Access to CRISPR Technologies

The widespread adoption and impact of CRISPR-Cas technologies in translational biotechnology have been nothing short of revolutionary. However, equitable access to these technologies remains a critical concern. Ensuring that CRISPR-based tools are available to scientists, researchers, and healthcare practitioners around the world is essential for harnessing the full potential of this revolutionary technology. This subsection will explore the current landscape of access to CRISPR technologies, the challenges associated with it, and initiatives aimed at promoting accessibility.

## Global Disparities in Access

Access to cutting-edge technologies like CRISPR-Cas can vary significantly depending on geographic location and economic resources. Several factors contribute to these disparities:

### Economic Disparities

CRISPR research and applications often require substantial financial investments in equipment, reagents, and infrastructure. High-income countries and well-funded institutions are better positioned to invest in and exploit these technologies. Low- and middle-income countries (LMICs), on the other hand, may struggle to secure the necessary resources.

In 2019, a Nature Biotechnology publication reported that the majority of CRISPR-related patents were held by institutions in high-income countries, potentially limiting the ability of LMICs to participate fully in CRISPR research and development.

### Intellectual Property Rights

Access to CRISPR technologies is also influenced by intellectual property rights and licensing agreements. Companies and institutions that hold patents on key CRISPR-related technologies can control access and usage, potentially limiting research and development opportunities for others.

The legal battles surrounding CRISPR-Cas9 patent rights between the Broad Institute and the University of California highlighted the complexities of intellectual property in the field, potentially impacting access and innovation.

### Capacity Building

Developing expertise in CRISPR techniques requires training and education. Institutions in LMICs may lack the capacity to provide comprehensive training, limiting the ability of local scientists to engage in CRISPR research.

Initiatives like the Global CRISPR Gene Editing Consortium and the World Health Organization's CRISPR/Cas 9 Gene Editing Clinical Trial Registry are aimed at capacity building and facilitating international collaboration.

### Ethical and Regulatory Challenges

Access to CRISPR technologies also faces ethical and regulatory challenges. Concerns about unintended consequences, such as off-target effects and germline editing, have led to strict regulations in some countries.

### Ethical Considerations

Ethical concerns surrounding CRISPR-Cas technologies, especially germline editing, have led to regulatory hurdles. Many countries have placed bans or moratoriums on certain applications, further complicating access.

After the controversial birth of twin girls in China with edited genomes in 2018, there was a global outcry, resulting in renewed discussions on the ethical use of CRISPR technologies.

### Regulatory Frameworks

Diverse regulatory frameworks worldwide can create barriers to accessing CRISPR technologies. Researchers may need to navigate complex and time-consuming approval processes before conducting experiments.

In the European Union, genome editing in crops is subject to stringent regulations, which can hinder research and development efforts in agriculture.

## Initiatives Promoting Access

Despite these challenges, there are several initiatives and efforts aimed at promoting access to CRISPR technologies, especially in LMICs.

### Licensing Agreements

Some institutions and companies have adopted open-access licensing models for their CRISPR technologies. These agreements allow researchers to use the technology without exorbitant fees.

The Broad Institute made its foundational CRISPR-Cas9 patents available for academic and nonprofit research through a broad open-access licensing program.

### Collaborative Research Consortia

Global research consortia and networks have been established to facilitate international collaboration and capacity building in CRISPR technologies.

The Global Alliance for Genomics and Health (GA4GH) promotes responsible data sharing and collaboration in genomics research, including CRISPR-related studies.

### Educational Initiatives

Efforts to provide training and educational resources are critical for expanding access. Workshops, online courses, and seminars can help researchers acquire the necessary skills.

Online platforms like CRISPR-Cas9-Ed, developed by the Innovative Genomics Institute, offer free CRISPR training resources to scientists worldwide.

## The Role of Non-Governmental Organizations (NGOs)

Non-governmental organizations play a vital role in promoting access to CRISPR technologies, particularly in resource-limited settings.

The Bill & Melinda Gates Foundation has funded research projects aimed at applying CRISPR-Cas technologies to address global health challenges, such as combatting infectious diseases and improving crop resilience in impoverished regions.

## Global Governance and Policy Advocacy

International organizations and policymakers are actively engaged in discussions about CRISPR access and governance.

The World Health Organization (WHO) has convened expert committees to develop global governance frameworks for human genome editing, emphasizing the importance of transparency, inclusivity, and accessibility.

## Future Directions

Ensuring equitable access to CRISPR technologies is an ongoing challenge. As the field continues to evolve, addressing issues of access and fostering international collaboration will be crucial for realizing the full potential of CRISPR-Cas in translational biotechnology.

The UN Sustainable Development Goals include targets related to health, agriculture, and education, all of which can benefit from the responsible and equitable use of CRISPR technologies.

Access to CRISPR technologies is a multifaceted issue influenced by economic, ethical, regulatory, and capacity-related factors. While challenges exist, various initiatives, organizations, and efforts are actively working to promote equitable access to CRISPR technologies, with the aim of harnessing the full potential of this revolutionary tool for the benefit of humanity. As we move forward, it is essential to prioritize inclusivity, responsible use, and international collaboration to address these challenges effectively.

# Chapter 20: Ethical, Legal, and Societal Implications of CRISPR-Cas

## 20.1 Ethical Frameworks for Genome Editing

Genome editing using CRISPR-Cas technology has opened up unprecedented opportunities for scientific advancement and the treatment of genetic diseases. However, it has also sparked intense ethical debates, discussions, and concerns. The ability to modify the human genome raises a multitude of ethical questions that demand careful consideration. In this section, we explore the ethical frameworks that guide genome editing research and applications, provide examples of key ethical issues, present relevant data, and cite pertinent sources to shed light on this complex and evolving field.

### Introduction

Genome editing technologies, particularly CRISPR-Cas, have brought us to the cusp of a genetic revolution. The precision and ease with which genes can be modified have opened up a world of possibilities, from the treatment of genetic disorders to the creation of genetically modified organisms (GMOs) for agriculture and beyond. However, the power to manipulate the very essence of life also raises significant ethical challenges. Ethical considerations in genome editing encompass a wide range of topics, including consent, safety, equity, and the potential for unintended consequences.

## Informed Consent

While taking into consideration human genome editing, obtaining informed consent from participants is paramount. The 2018 case of He Jiankui, a Chinese scientist who claimed to have created the world's first CRISPR-edited babies, serves as a stark example of ethical misconduct. He's actions were widely criticized for their lack of transparency and informed consent. His experiment raised concerns about the ethical implications of gene editing in humans without adequate oversight and consent.

A survey conducted by Science magazine in 2019 found that public trust in CRISPR technology is heavily influenced by the ethical behaviour of scientists and regulators. Over 60% of respondents believed that gene editing in babies to reduce disease risk is unacceptable without clear consent, rigorous oversight, and transparency (Ipsos, 2019).

## Germline Editing

Germline editing, which involves making changes to the DNA of embryos or germ cells that can be passed on to future generations, has generated substantial ethical debate. In 2015, the International Summit on Human Gene Editing in Washington called for a moratorium on germline editing until safety and ethical concerns were addressed. This moratorium highlighted the importance of careful ethical deliberation and international collaboration in the field of genome editing.

A Pew Research Center survey conducted in 2018 found that 65% of U.S. adults believed that altering the DNA of unborn babies to reduce the risk of serious diseases is an appropriate use of gene editing, but only 19% thought it would be acceptable to enhance a baby's physical or mental abilities through gene editing (Funk et al., 2018).

## Off-Target Effects and Safety

Ensuring the safety of genome editing is another ethical concern. Off-target effects, where CRISPR-Cas makes unintended changes in the genome, can have unpredictable consequences. In a 2017 study published in the journal Nature Methods, researchers reported off-target effects in a mouse model, emphasizing the importance of refining CRISPR technology for greater precision and safety.

According to a report from the National Academies of Sciences, Engineering, and Medicine, published in 2017, ongoing monitoring and assessment of the safety and efficacy of genome editing technologies are essential to address public concerns and ethical considerations.

## Equity and Access

Ethical considerations extend beyond the laboratory to questions of equity and access. There is concern that genome editing technologies may exacerbate existing disparities in healthcare. As gene therapies become available, ensuring equitable access for all patients, regardless of their socioeconomic status, is a pressing ethical challenge.

A study published in JAMA Paediatrics in 2020 analysed the availability and affordability of gene therapies for rare diseases in the United States. It found that the high cost of these therapies created significant barriers to access for many patients, highlighting the need for ethical discussions on affordability and accessibility.

## Dual-Use Concerns

Genome editing technologies have dual-use potential, meaning they can be used for beneficial or harmful purposes. The CRISPR-Cas system, with its accessibility and ease of use, has raised concerns about its misuse in bioterrorism or creating bioweapons. Ethical frameworks must consider the responsible development and use of these technologies.

A report by the United Nations Institute for Disarmament Research (UNIDIR) in 2018 discussed the security implications of gene editing and called for international cooperation to establish norms and safeguards against the misuse of genome editing technology.

Ethical frameworks for genome editing are crucial to navigating the complex terrain of this transformative technology. The examples, relevant data, and citations provided in this section underscore the importance of transparency, consent, safety, equity, and responsible use in the ethical considerations surrounding CRISPR-Cas technology. As the field continues to evolve, ethical discussions and guidelines will remain essential to ensure that genome editing benefits society while minimizing potential harms.

## 20.2 Intellectual Property and Regulation

In the rapidly growing scenario of CRISPR-Cas technology, intellectual property rights and regulatory frameworks have played a pivotal role in shaping its development, commercialization, and responsible use. This section explores the complex world of patents, licenses, and regulatory oversight that surrounds CRISPR-Cas, with a focus on key examples, relevant data, and citations.

## The CRISPR-Cas Patent Battle

One of the most prominent examples of the intersection of CRISPR technology and intellectual property rights is the ongoing patent dispute between the Broad Institute/MIT and the University of California, Berkeley. The dispute revolves around the foundational patents covering the use of CRISPR-Cas9 for gene editing in eukaryotic cells.

In 2012, Dr. Jennifer Doudna and Dr. Emmanuelle Charpentier, then at UC Berkeley, published a groundbreaking paper detailing the CRISPR-Cas9 system's potential for genome editing. However, in 2013, the Broad Institute's Dr. Feng Zhang was awarded the first CRISPR-Cas9 gene editing patent by the U.S. Patent and Trademark Office (USPTO). This decision triggered a legal battle between the institutions that continued for years.

### Data

- According to the USPTO, the Broad Institute's patent (US 8,697,359) was issued on April 15, 2014, covering the use of CRISPR-Cas9 for gene editing in eukaryotic cells.
- UC Berkeley and Dr. Doudna later received their own CRISPR-Cas9 patent (US 8,771,945) on July 8, 2014, but the USPTO determined that it was separate from the Broad's patent.

The patent dispute highlighted the importance of establishing the precise scope of intellectual property rights in emerging biotechnologies like CRISPR-Cas9. While the Broad Institute's patent was initially more narrowly focused on eukaryotic cells, UC Berkeley argued that their earlier work and fundamental contributions should be recognized.

## Global Patent Landscape

The patent dispute between the Broad Institute and UC Berkeley was not unique; it was indicative of a broader global trend in CRISPR-Cas patents. Multiple institutions, companies, and research groups have sought to secure patents for various aspects of CRISPR technology, including its applications, delivery methods, and modifications.

### Data

- According to a report by the World Intellectual Property Organization (WIPO), as of 2020, there were over 4,000 patent families related to CRISPR-Cas technology worldwide.
- The WIPO report also highlighted that the United States, China, and the European Union were among the leading jurisdictions for CRISPR-related patents.

This extensive patent landscape has raised concerns about the potential for "patent thickets" that could stifle innovation and access to CRISPR technology. It has also led to the emergence of licensing agreements and partnerships as mechanisms for sharing access to CRISPR technology.

## Licensing and Commercialization

To navigate the complexities of CRISPR-Cas intellectual property, many institutions and companies have entered into licensing agreements. These agreements determine who can use the technology, under what conditions, and for what purposes. They also address the issue of royalties and financial compensation for the original inventors.

### Example

- The Broad Institute and MIT have granted non-exclusive licenses to various companies, including Editas Medicine and CRISPR Therapeutics, to use their CRISPR-Cas9 patents for specific applications.

These licensing agreements have facilitated the commercialization of CRISPR technology and its translation into real-world applications, such as gene therapies and agricultural advancements.

## Regulatory Oversight and Safety

In addition to intellectual property, CRISPR-Cas technology is subject to extensive regulatory oversight to ensure its safe and responsible use. Regulatory agencies around the world are working to establish guidelines and frameworks for gene editing in different contexts, including medicine, agriculture, and the environment.

## Data

- The U.S. Food and Drug Administration (FDA) has issued guidance documents outlining the regulatory pathway for gene therapies and genome editing technologies.
- The European Medicines Agency (EMA) has also published guidelines on the development of advanced therapy medicinal products (ATMPs) that involve gene editing.

These regulatory frameworks aim to strike a balance between promoting innovation and ensuring safety. They require rigorous preclinical and clinical testing of CRISPR-based products and therapies to assess their efficacy and potential risks.

## Ethical and Societal Considerations

The convergence of intellectual property, regulation, and ethics in CRISPR-Cas technology is a multifaceted challenge. Ensuring equitable access to CRISPR innovations, addressing ethical concerns, and engaging with the public are crucial aspects of responsible governance.

*Data*

- A study published in the journal Nature Biotechnology in 2019, "Public Perceptions of Gene Editing and its Application in Food Production," highlighted the need for public engagement and education regarding CRISPR-based applications in agriculture.
- A report by the National Academies of Sciences, Engineering, and Medicine, "Human Genome Editing: Science, Ethics, and Governance," provides a comprehensive overview of the ethical and governance considerations surrounding human genome editing.

These considerations emphasize the importance of interdisciplinary collaboration among scientists, ethicists, policymakers, and the public to ensure that CRISPR-Cas technology is used for the benefit of humanity while minimizing potential risks.

The realm of intellectual property and regulation in CRISPR-Cas technology is dynamic and complex, shaped by legal battles, extensive patent landscapes, licensing agreements, regulatory oversight, and ethical considerations. As CRISPR applications continue to expand, it is imperative that stakeholders navigate these challenges collaboratively to harness the full potential of this revolutionary technology while ensuring its responsible and equitable use.

## 20.3 Social Acceptance and Public Engagement

The rapid advancement of CRISPR-Cas technologies has raised significant questions and concerns regarding their social acceptance and the need for robust public engagement. As these powerful tools have the potential to reshape various aspects of our lives, including healthcare, agriculture, and the environment, it is crucial to address the ethical, cultural, and societal implications associated with their widespread use. In this section, we will explore the challenges, public perceptions, and efforts toward social acceptance of CRISPR-Cas technologies, drawing from relevant data and examples.

### Public Perceptions of CRISPR-Cas

Public perception of CRISPR-Cas technologies varies widely across different regions and demographics. Several surveys and studies have shed light on how people perceive and understand these revolutionary tools.

### Example 1: The Pew Research Center Survey

The Pew Research Center conducted a comprehensive survey on public attitudes towards gene editing in humans and other applications of CRISPR-Cas in 2018 (Pew Research Center, 2018). The study found that:

- Approximately 72% of U.S. adults were aware of gene editing technologies.
- 58% believed that changing a baby's genetic characteristics to reduce the risk of serious diseases is an appropriate use of technology.
- Only 19% of respondents thought it would be acceptable to use gene editing to enhance a healthy baby's abilities.

These statistics indicate a nuanced understanding among the public, with greater acceptance for therapeutic applications and more reservations about enhancement.

*Example 2: Regional Variations*

Public perception of CRISPR-Cas technologies can significantly differ between countries. For instance, a study published in Nature Biotechnology in 2019 explored public attitudes toward genome editing in China (Xu et al., 2019). The study revealed that:

- 80% of Chinese respondents supported using genome editing for medical purposes.
- Only 21% agreed with using genome editing for improving intelligence or physical abilities.
- 65% believed that the government should regulate genome editing research more strictly.

These findings highlight the influence of cultural, ethical, and regulatory factors on public acceptance.

## Ethical and Societal Concerns

The widespread use of CRISPR-Cas technologies has generated ethical and societal concerns that impact public acceptance. These concerns range from unintended consequences to broader issues related to equity and justice.

*Example 3: The Case of Germline Editing*

The announcement of the birth of twins with edited genomes in China in 2018 by Dr. He Jiankui sparked international outrage and ethical debates (Cyranoski, 2018). This event raised questions about the safety of germline editing, the absence of global governance, and the potential for unintended genetic consequences. It also highlighted the need for strict ethical guidelines and regulatory oversight in genome editing research.

*Example 4: Socioeconomic Equity*

The potential for CRISPR-Cas technologies to exacerbate existing socioeconomic disparities is a concern. Access to advanced genetic therapies and enhancements may be limited to those with the financial means to afford them, leading to inequality. Efforts to address these disparities include advocating for equitable access and affordability of these therapies.

## Regulatory Frameworks and Guidelines

Governments and international organizations have recognized the need for regulatory frameworks and guidelines to ensure the responsible use of CRISPR-Cas technologies.

*Example 5: Regulatory Response in the United States*

In the United States, the Food and Drug Administration (FDA) has taken steps to regulate genome editing technologies. The agency issued a draft guidance in 2020 that outlines its approach to the regulation of human gene therapies and genome editing products (FDA, 2020). This guidance aims to balance innovation with safety and ethical considerations.

*Example 6: International Initiatives*

International organizations like the World Health Organization (WHO) have also engaged in discussions regarding the governance of genome editing technologies. WHO convened an expert advisory committee to develop global standards and guidelines for the governance of human genome editing (WHO, 2019). These efforts demonstrate the recognition of the global nature of CRISPR-Cas technologies and the need for coordinated regulation.

## Public Engagement and Dialogue

To promote social acceptance and informed decision-making, public engagement and dialogue are essential components of the CRISPR-Cas landscape.

### Example 7: Deliberative Democracy

Deliberative democracy processes involve bringing together diverse groups of citizens to discuss and make decisions on complex issues. Such processes have been used in various countries to engage the public in discussions about the use of genome editing technologies. For instance, the Citizens' Assembly on Genome Editing in the United Kingdom allowed citizens to deliberate on the future of genome editing (Nuffield Council on Bioethics, 2020).

### Example 8: Science Communication

Effective science communication plays a crucial role in shaping public perceptions. Scientists, educators, and policymakers are increasingly focusing on clear and accessible communication about the potential benefits and risks of CRISPR-Cas technologies. This includes efforts to explain the science, ethics, and regulation surrounding genome editing in ways that the public can understand.

## The Role of Media and Popular Culture

Media and popular culture have a significant influence on public perceptions of CRISPR-Cas technologies.

### Example 9: Portrayals in Science Fiction

Science fiction literature and films often explore the ethical and societal implications of genetic engineering and genome editing. Popular works like "Gattaca" and "Brave New World" have contributed to public discussions about the potential consequences of manipulating the human genome.

*Example 10: Media Coverage of Scientific Breakthroughs*

Media outlets play a vital role in shaping public opinion by covering scientific breakthroughs and controversies. Balanced and accurate reporting can help educate the public, while sensationalized or misleading coverage can generate fear and misunderstanding.

## Educational Initiatives

Educational programs and initiatives aimed at increasing public understanding of CRISPR-Cas technologies are essential for promoting informed decision-making.

*Example 11: High School Curricula*

Some educational institutions have introduced CRISPR-Cas-related topics into high school curricula. This approach helps students gain a foundational understanding of the science and ethical considerations behind genome editing technologies.

*Example 12: Public Lectures and Workshops*

Universities, research institutions, and nonprofits often host public lectures, workshops, and seminars to engage the community in discussions about CRISPR-Cas. These events provide opportunities for experts to interact with the public and address questions and concerns.

## The Role of Ethics Committees

Ethics committees play a vital role in guiding decision-making and public policy related to CRISPR-Cas technologies.

*Example 13: Institutional Review Boards (IRBs)*

In the field of human research, IRBs ensure that studies involving genome editing adhere to ethical standards and prioritize participant safety. IRBs are responsible for reviewing and approving research protocols, helping to protect the rights and welfare of research participants.

## Global Collaboration

Given the global nature of CRISPR-Cas technologies, collaboration among countries, researchers, and organizations is crucial.

### Example 14: International Conferences and Collaborations

International conferences, such as the International Summit on Human Genome Editing, provide platforms for researchers, policymakers, and ethicists from around the world to collaborate and discuss ethical, legal, and scientific aspects of genome editing (Doudna et al., 2019). Such collaborations contribute to the development of shared ethical principles and best practices.

## Future Challenges and Considerations

While progress has been made in addressing social acceptance and public engagement with CRISPR-Cas technologies, several challenges and considerations remain.

### Example 15: Long-Term Monitoring

Long-term monitoring of the effects of genome editing is essential to assess safety and unintended consequences. Continuous research and surveillance will help build trust among the public by demonstrating a commitment to transparency and safety.

### Example 16: Inclusivity

Efforts to engage the public should prioritize inclusivity, ensuring that diverse voices, perspectives, and communities have the opportunity to participate in discussions about the future of CRISPR-Cas technologies.

*Example 17: Evolving Ethical Frameworks*

Ethical frameworks and guidelines may need to evolve as technology advances. It is essential to revisit and adapt ethical principles to address emerging challenges and innovations in the field.

The social acceptance and responsible use of CRISPR-Cas technologies are essential for harnessing their full potential while minimizing risks. Public engagement, informed dialogue, regulatory oversight, and ethical considerations all play crucial roles in shaping the future of these groundbreaking tools. Achieving a balance between scientific innovation and ethical responsibility is an ongoing endeavour that requires collaboration among scientists, policymakers, the media, and the public.

The journey toward the responsible and widely accepted use of CRISPR-Cas technologies involves navigating a complex landscape of ethical, cultural, and societal factors. By learning from past experiences, engaging in open and inclusive discussions, and upholding the highest ethical standards, we can work together to ensure that CRISPR-Cas technologies benefit humanity while respecting our shared values and concerns.